The Changing Face of Alterity

Media Philosophy

Series Editors:

Eleni Ikoniadou, Lecturer in Media and Cultural Studies at the London Graduate School and the School of Performance and Screen Studies, Kingston University
Scott Wilson, Professor of Cultural Theory at the London Graduate School and the School of Performance and Screen Studies, Kingston University

The *Media Philosophy* series seeks to transform thinking about media by inciting a turn toward accounting for their autonomy and 'eventness', for machine agency, and for the new modalities of thought and experience that they enable. The series showcases the 'transcontinental' work of established and emerging thinkers whose work engages with questions about the reshuffling of subjectivity, of temporality, of perceptions and of relations vis-à-vis computation, automation, and digitalisation as the current twenty-first-century conditions of life and thought. The books in this series understand media as a vehicle for transformation, as affective, unpredictable, and non-linear, and move past its consistent misconception as pure matter-of-fact actuality.

For Media Philosophy, it is not simply a question of bringing philosophy to bear on an area usually considered an object of sociological or historical concern, but of looking at how developments in media and technology pose profound questions for philosophy and conceptions of knowledge, being, intelligence, information, the body, aesthetics, war, and death. At the same time, media and philosophy are not viewed as reducible to each other's internal concerns and constraints and thus it is never merely a matter of formulating a philosophy of the media; rather the series creates a space for the reciprocal contagion of ideas between the disciplines and the generation of new mutations from their transversals. With their affects cutting across creative processes, ethico-aesthetic experimentations and biotechnological assemblages, the unfolding media events of our age provide different points of intervention for thought, necessarily embedded as ever in the medium of its technical support, to continually re-invent itself and the world.

"*The new automatism is worthless in itself if it is not put to the service of a powerful, obscure, condensed will to art, aspiring to deploy itself through involuntary movements which none the less do not restrict it.*"

Eleni Ikoniadou and Scott Wilson

Titles in the Series

Software Theory, Federica Frabetti
Media After Kittler, edited by Eleni Ikoniadou & Scott Wilson
Chronopoetics: The temporal Being and Operativity of Technological Media, Wolfgang Ernst, translated by Anthony Enns
The Changing Face of Alterity: Communication, Technology and Other Subjects, edited by David J. Gunkel, Ciro Marcondes Filho & Dieter Mersch
Algorithmic Catastrophe: On the Contingency and Necessity of Technical Systems, Yuk Hui

The Changing Face of Alterity

Communication, Technology, and Other Subjects

Edited by
David J. Gunkel, Ciro Marcondes Filho and Dieter Mersch

ROWMAN &
LITTLEFIELD
INTERNATIONAL
London • New York

Published by Rowman & Littlefield International, Ltd.
Unit A, Whitacre Mews, 26-34 Stannary Street, London SE11 4AB
www.rowmaninternational.com

Rowman & Littlefield International, Ltd. is an affiliate of Rowman & Littlefield
4501 Forbes Boulevard, Suite 200, Lanham, Maryland 20706, USA
With additional offices in Boulder, New York, Toronto (Canada),
and Plymouth (UK)
www.rowman.com

British Library Cataloguing in Publication Data
A catalogue record for this book is available from the British Library

ISBN: HB 978-1-7834-8869-8
PB 978-1-7834-8870-4

Library of Congress Cataloging-in-Publication Data
Names: Gunkel, David J., editor.
Title: The changing face of alterity : communication, technology, and other subjects / edited by David J. Gunkel, Ciro Marcondes Filho & Dieter Mersch.
Description: London : Rowman & Littlefield International, 2016. | Series: Media philosophy | Includes bibliographical references and index.
Identifiers: LCCN 2016033798 (print) | LCCN 2016035273 (ebook) | ISBN 9781783488698 (cloth : alk. paper) | ISBN 9781783488704 (pbk. : alk. paper) | ISBN 9781783488711 (Electronic)
Subjects: LCSH: Other (Philosophy) | Communication. | Technology.
Classification: LCC BD460.O74 C425 2016 (print) | LCC BD460.O74 (ebook) | DDC 302.201--dc23
LC record available at https://lccn.loc.gov/2016033798

∞™ The paper used in this publication meets the minimum requirements of American National Standard for Information Sciences—Permanence of Paper for Printed Library Materials, ANSI/NISO Z39.48-1992.

Printed in the United States of America

Contents

Acknowledgments

This book began at breakfast during the second *Encontro Nacional da Rede de Grupos de Pesquisa em Comunicação* [The National Meeting of the Network of Research Groups in Communication], which was convened in Natal, Brazil in November of 2013. Present at that first "planning session" were the three editors whose names appear on the spin of this book. But also present was an Other who is solely responsible for getting this project started and off the ground—Lauren Ferreira Colvara. It was her vision, her tireless effort, and her ongoing support that made this book, and the international meetings in São Paulo (October 2014) and Zürich (January 2015) that served as its incubator, possible in the first place. It is for this reason, that we dedicate this book to her with our gratitude—Thank you | Denke sehr | Muito obrigado.

We would also like to recognize and thank the organizations that provided funding, facilities, and support for the international meetings that occurred in Brazil and Switzerland: Coordenação do Aperfeiçoamento do Pessoal de Ensino Superior (CAPES), Fundação de Amparo à Pesquisa do Estado de São Paulo (FAPESP), the University of São Paulo, and the Institute for Critical Theory of Zürich University of the Arts.

Introduction

David J. Gunkel, Ciro Marcondes Filho and Dieter Mersch

The figure of the Other is fundamental to both the theory and practice of communication. Communication, which is, as James Carey (1992, 15) has argued, most commonly defined as the act of sending or imparting information to others, is only possible in the face of others. And all acts of communication seek to respond to or address themselves to an Other—whether the "other person" in face-to-face interpersonal interactions; the "audience" in public speaking, mass communication, and broadcasting; the "consumer" in marketing communications and public relations; or the "user" of web content and Internet applications. Consequently, no matter how it is structured or conceptualized, communication is involved with addressing others and dealing with the ontological, epistemological, and ethical questions of otherness or alterity. Within the discipline of Communication Studies, this Other is almost always assumed to be another human being. As one of the "human sciences," Communication is organized around the human subject and the conditions of human intersubjectivity. The face of the Other, in other words, has been and remains exclusively human. This widely accepted and often unquestioned precondition is now confronted with and challenged by recent innovations in information and communication technology (ICT), where the appearance of and encounter with the Other is mediated by various forms of computer interface, hidden behind the virtual mask of an avatar, or even replaced by faceless algorithms and the artifice of social robots.

This book seeks to investigate the opportunities and challenges that now face us with current reconfigurations of the other. Drawing on the research and expertise of scholars from three continents (Europe, North America, and South America), the book assembles critical essays that investigate the many facets of otherness as it is experienced in contemporary techno-culture. This effort is directly communicated by the book's title, *The Changing Face of Alterity*, which indicates the advent of different configurations of otherness made possible by technologies that appear, in one way or another, to reformulate or reconfigure the face of the other. This title is followed by a subtitle that lists the three subjects that are considered by the book's contributors: the phenomenon of *communication* as the site for an appearance of and experience with alterity; the impact and influence of recent innovations in *technology* that can mask, reconfigure, or even simulate different facets of the other; and *other subjects* like the fundamental conditions of human intersubjectivity, the experience of being subjected to the face of another (and other kinds of others), and the subject of "face" as a philosophical concept.

The ten essays collected in the book provide a range of different responses to this challenge, extending from, on the one hand, efforts to redefine and rehabilitate the concept of human exceptionalism within the human sciences in general and Communication Studies in particular to, on the other hand, proposals for a posthuman theory and practice of communication that takes seriously the impact and importance of other forms of otherness, like animals and machines. Consequently the main objective of the book is not to resolve the question concerning alterity once and for all but to initiate informed, critical debate by describing what needs to be asked about and investigated in the first place. This is not, it is important to point out, a cop out; it is the proper domain and necessary work of philosophy. Philosophers as varied as Martin Heidegger (1962), Daniel Dennett (1996), George Edward Moore (2005), and Slavoj Žižek (2006) have all, at one time or another, argued that the principal objective of philosophy is not to supply answers to difficult questions but to formulate and develop the kinds of innovative questions that should be asked. As Dennett (1996, vii) has explained: "I am a philosopher, not a scientist, and we philosophers are better at questions than answers. I haven't begun by insulting myself and my discipline, in spite of first appearances. Finding better questions to ask, and breaking old habits and traditions of asking, is a very difficult part of the grand human project of understanding ourselves and our world."

1. THE FACE OF THE OTHER

The book is divided into three sections. The first, "The Face of the Other," begins by investigating the impact recent innovations in ICT have on the ontological, epistemological, and moral aspects of *the Other* and the philosophical concepts of *face* and *alterity*. It begins with Dieter Mersch's critical examination of faces, masks, and avatars as three different modalities of "facing," each of which involves different kinds of relationships. While the avatar (or the robot) serves as a surface or a screen on which we project human features, much like we project personal characteristics on props or puppets, it only represents a mask, the symbolic face of the dead or of the animal and ghosts. It therefore, Mersch argues, only mediates the Other and by doing so presupposes a human agent as its implicit substratum. We can, Mersch continues, only encounter the face of the Other as a vulnerable and fragile countenance that is, at the same time and according to Levinas, able to transcend its visual appearance as the trace of divinity. This duplicity in the face of the face can be explicitly demarcated by way of the two different German notions of *Gesicht* [face = identity] and *Antlitz* [countenance = alterity]. The face, Mersch argues, conveys a primordial ethical meaning which does not hold (or at least not in the same way) for the relationship with masks and *a fortiori* with avatars (or robots). The first chapter therefore argues that the relationship with faces exceeds any relation that one might have with artifacts and, for this reason, serves as a proper model for our relationship with virtual objects such as avatars or robots. Hence, there is a fundamental difference between the relationship we have with things and the relationship we have with other humans, which indicates, at the same time, a certain primacy of the intersubjective realm or the social. We are, Mersch concludes, always already infused with the social and the Other before we are able to relate to artificial objects, so that we can, by no means, act on the assumption of a similitude or symmetry like many of the posthumanists argue. Instead, Mersch makes a case for different *modes of relatedness* between humans, robots, and avatars; and thus on fundamental differences in both perception and the way we respond to others and their artificial substitutes. And these differences correspond with the line that separates responsivity and intentionality, on the one hand, and passivity and activity, on the other, whereby the former constitutes and informs the latter with an *a priori* ethical obligation and impact.

The second chapter, "Digital Exchanges: Ghosts and Gifts" by Mira Fliescher, tries to outline an ethics of the gift of facing in the digital age by reconsidering Jacques Derrida's and Emmanuel Levinas's ethics of alterity. The digital age, Fliescher argues, is marked by the task of sharing, by gaining a face (a social face) by sharing information, making comments, and clicking "like" and "dislike." In this activity, Fliescher argues, we do not only share information about ourselves but also about other persons. Though the attempt to protect privacy is growing, our activities on social media are based on the effort to make oneself public. Most of these things happen without the real presence of the other, as the subjectivities and acts in these social networks are governed by the needs and restrictions of the algorithms running them. While this exchange produces valuable information for the economic purpose of maximizing profit (i.e., cheap data for market research), the question that remains to be asked is "What kind of ethics does this sociality enable or require?" When the social in the digital age seems to be structured by an economics of exchange and on giving back (since every post asks for a formalized response), of counting and quantification (the number of friends, comments, or shares), and of an absence of the other that is shaped by algorithms, then this is not, Fliescher argues, the same as "writing" in the sense that Derrida develops. The infiniteness of the algorithms is not infinity (one of Levinas's keywords) as it tends more toward a repetition of the same than to infinite *différance*. And if there is something like a "digital responsiveness," it does not occur in the presence of the face of the Other nor does it regard the bareness of the other. It is, Fliescher contends, nothing but a reply to a formal algorithm, and therefore, there is not even the presence of the ghost of the Other. What, then, is sharing, gifting, facing, or alterity? The essay follows this questions in order to diagnose the loss of and the need for an aesthetics of the Other that gives a face of/for the Other. And it is for this reason, that Fliescher considers Levinas's concept of *substitution* as well as Derridas concept of the *gift*. In doing so, Fliescher opens up a new critique of social media as well as formulates a deconstructive ethics of alterity. She shows that, with regard to ethics, it is not wise to transfer the concept of writing to social networks or digital media. Since writing is not, she concludes, the same as code or a program. Therefore, in a turn against the repudiation of aesthesis and a general disregard of the essential role of materiality in Derrida's and Levinas's writings, she argues for the need to develop an aesthesis of the Other that is carried out by the materiality of mediality substituting for the presence of the Other. This gift of an impossible substitution in materiality happens before there is any intentional consciousness in play, and it is a necessity that cannot be fulfilled by communication run by algorithms and programs.

The final chapter in the first section is Jörg Sternagel's "Performative Modalities of Otherness." Within correlations between being and alterity, sense and meaning, are situated the ethical and the performative. Uniting the subjectivity of perceiving—intentionally aiming at an object—with the objectivity of expressing and creating, such as occurs in language, a play on stage or in film, or in a dance, reflection begin on an ontological order of corporeal, linguistic, and artistic gestures, where the encounter with another is in a cultural whole, in which she or he is present and illuminated by this whole, as a text would be by its context. From this starting point, Sternagel develops a phenomenology of technics and media that is grounded on its attention to detail and its emphasis on the relationship between the human being and technology or the human-technology relationship. Its task, Sternagel argues, is to discover the various structural features of those relations—its material, tactile, and tangible dimensions and its impact on our perceptions. Its objective is to discover the multiple ways in which my body interacts with the world surrounding it and with its environment by means of technologies. The chapter, therefore, investigates the ethical dimensions of the performative, which unfold and enfold with regard to the mediality and *mise-en-scène* of digital performances focused on renderings as mediation between bodies and screens, where the manifestation of the human form gradually becomes invisible, while its rendering successively becomes visible. The point of contact for Sternagel's analysis is the *avatar*, but not the avatar of computer games or social media. He is instead concerned with the *becoming avatar* of the human actors in digital cinema. Unlike traditional forms of filmmaking, which remain as Lev Manovich (1995) argues an "indexical art," digital cinema, in general, and CGI or computer-generated imagery, in particular, transforms the face of the actor through green screen techniques that strip away backgrounds and bodies and reconfigures his/her facets and features. In doing so, the actor, the one who had once stood behind and animated the mask in classical theater, gradually surrenders to a digital fragmentation and encoding that makes the other appear otherwise. Thus the spectator of digital cinema watches a digitally processed performance and becomes involved with another—the "alienness" of an avatar—in its opaque impenetrability that fails to establish any true relationship.

2. FACING OTHERS

The second section, "Facing Others," focuses on the ontological, epistemological, and moral situation of communication that is the condition for possibility of exposure to and an encounter with others. It begins with a chapter from Ciro Marcondes Filho, who asserts that communication is only possible with alterity—not just any kind of alterity but a radical one that remains alien in its *incapturability* or its eternal fugue. The experience of the other or his/her impenetrability, however, is not enough. It is also necessary that the other does not appear by chance or just accidentally, but as an expression, or better as a saying, that his/her/its appearance in front of me makes me think and changes my situation by creating meaning. In this sense, Marcondes Filho argues, alterity is different from otherness, which can be seen in our relationship with animals, robots, and nature itself. For them, even in their strangeness before me, this dimension, their communicability, is not present. Animals and pets, Marcondes Filho maintains, do not communicate complex processes, although they may indicate what they feel, what they want, or how they want it, or where they want to go. Their verbal understanding is understandably limited; they do not capture the broader and more abstract context. Machines do not communicate either. They can talk, build contexts, and give unpredictable responses, but they lack the subtle indeterminacy of the human, especially in Eros, where the one who is closest to me, in an absolute intimate relationship, can flee back into his/her inaccessibility as a Being that teases me and makes an opening in myself to becoming. With technology, Marcondes Filho concludes, the face of alterity has changed. We do not recognize the robot as having communicability; its otherness is not experienced as an enlightened mystery. Computer-mediated human relations, by contrast, are able to reproduce the radical alterity that Eros would attribute only to direct and in-person interactions.

In chapter 5, Maurício Liesen both extends and elaborates this conclusion, performing a media critique that argues that there is an opposition between the experience of play in computer games and the genuine experience of alterity. According to Liesen, play can be understood as a *threshold-experience*, a space situated "in-between" and always perceived as a frame (an "as-if"). Alterity, by contrast, is characterized as a *limit-experience* or an experience of the irreparable, which opens up and exposes the ethical relationship with an Other. Through this theoretical distinction between the medial function of play and the communicational function of alterity, Liesen criticizes the use of the concept of alterity in the characterization of experiences with digital technology. For him, the play logic, based on interaction, is substantially different from the responsiveness of alterity. Furthermore, he argues that the underlying logic behind social media is also not communication, but rather digital (mathematical) play. For Liesen, then, the distinction between play and the experience of alterity provides a useful tool to criticize the very idea of a communication society based on digital devices. Thus, the chapter also constitutes an apologia for the specificity of assigning the concept of alterity to face-to-face human communication.

In the sixth chapter, Alex Galeno investigates the social and political dimensions of these alterations in alterity. He begins by asking whether it is possible, in the contemporary digital era, to think of communication without a connection between the subject of communication and the media of subjectivity and the modes of visibility that are produced by this relationship. For Galeno, the important philosophical question is not only "what are we?" but "*where* are we?" and "*how* are we?" when we are interacting with our technical devices. For this reason, Galeno suggests that concepts like connectivity, faciality, alterity, and spatiality need to be understood as ontological and epistemological categories of the phenomenon and technology of communication. When interacting online, for instance, we are inexorably connected to the world of other things and other human beings. If it is true that with the advent of virtual communication, faces often become invisible or metamorphose into the carefully constructed and controlled interface of avatars and profiles, then we might ask what becomes of the concepts of the "public sphere" and of social and political action. Very different from the Luddites, who sabotaged technocratic formations by intentionally breaking the machines, we are currently, Galeno writes, witnessing other forms of revolt, like that of the Rolezinho, a social movement of young people from São Paulo's periphery, which mainly occurred in 2013. How, Galeno asks, do we classify this new form of social and political opposition? Especially because these groups have their meetings and demonstrations at the center of hegemonic control, for example, in the shopping centers, the malls, and the Crystal Palaces of consumption. Galeno carefully considers the case of the Rolezinhos, who use the technology of the Internet, Facebook, and mobile devices as prostheses of sociability to form alternative, oppositional communities engaged in what initially appears (from the outside, at least) to be the will to nothingness.

In the final chapter of this section, Ann Hetzel Gunkel begins to question and complicate these insights, especially the way we typically differentiate the "real" from the "virtual." She does this through a reflection on a research project—*Digital Diversities: Social Media and Intercultural Experience*—carried out between 2012 and 2014 at the Jagiellonian University in Kraków, Poland. Initially, Hetzel Gunkel points out how difficult it is to adopt a term such as "digital native" for a whole generation of young people whose "natural" speech would appear to be contextualized by computer technology. In reality, she argues, things are not so simple, primarily because these technologies do not always seek out the Other but result in the Same. The research for the study was based on interviews, where respondents revealed a sincere but somewhat confused formulation of the "real" as opposed to the "virtual." According to Hetzel Gunkel, the rhetoric of the "real" shapes the understanding and explanation of young people both online and offline, making the real a hegemonic concept. There is, then, a kind of Platonic idea in the thinking and reason of these young people—one that distinguishes between the real thing, which is viewed positively, and its phenomenal appearances or virtuality, which is often characterized in negative terms. Students, therefore, incorporate ancient and medieval concepts of technology and metaphysics, without ever questioning them. In the end, Hetzel Gunkel suggests that the apparent confusion of the student respondents is derived from the very definition of communication, which is typically based on the standard model of senders and receivers exchanging information. This may, she argues, be an oversimplification of Levinas, who wrote that the face-to-face should be restored as a genuine function of communicability. In response to this, Hetzel Gunkel suggests that it is in the work of Slavoj Žižek that we might find an alternative formulation, where the real is not some reality standing behind the virtual simulation but is an original emptiness that already makes reality incomplete and inconsistent. It is, therefore, in the face of the virtual that we may begin to see how our "real life" has already been virtual all along.

3. INTERFACES AND OTHER FACES

The third and final section, "Interfaces and Other Faces," picks up where "Facing Others" leaves off. Whereas previous chapters sought to withhold or at least resist extending alterity to other entities like computers, robots, artificial intelligence, and algorithms, the chapters in this section ask whether and to what extent the machines of our own making are able to possess face and be considered Other. This final section begins with Tales Tomaz's investigation of alterity and technology in the work of Martin Heidegger. Although Heidegger famously pursued the question concerning technology, his philosophy is often criticized for its lack of a serious engagement with others questions, specifically those moral questions asked in the face of Others. As Silvia Benso (2000, 127) has described it, following the insights of Levinas, "There is no ethics in Heidegger, at least according to the most common reading." Tomaz demonstrates that this "common reading" is not only inaccurate but that it is in and by "The Question Concerning Technology" that Heidegger's thinking of the Other comes to be revealed. Against the current of Heideggarian orthodoxy, Tomaz argues that Heidegger did in fact make a place for the Other and that this place was coextensive with the mode of the revealing of being that is made possible by the *enframing* of modern technology. Because of this, Tomaz argues, we are confronted with two important consequences. On the one hand, human beings relate to themselves and each other as mere things, that kind of standing reserve whereby *Dasein* is reduced to its mere social roles and functions. But, and on the other hand, other things, like machines and technology, which are typically considered mere instruments of human action, are also other beings that are capable—in theory, at least—of occupying the same positions as humans in the age of technology. Finally, Tomaz concludes (following Heidegger but not necessarily using his terminology), this *being cyborg*, which affects both the human being and our technologies, is not a choice or something that could be avoided. It is a *fait accompli*. Resistance is futile.

If Tomaz's unorthodox reading of Heidegger opens up the possibility of another way to situate and consider alterity in and of technology then the opportunities and challenges of this innovation are pursued in the chapters that follow. In chapter 9, "Alterity Ex Machina," Mark Coeckelbergh investigates the machine as other. He begins by demonstrating how human beings have always needed and cannot do without others.

The "human," Coeckelbergh argues, is only able to be defined and characterized by being differentiated from others, most notably non-human animals and machines—those two "others" that René Descartes had combined together in a single concept, the *bête-machine*. According to Coeckelbergh, our philosophical anthropologies and ontologies use these others (or instrumentalize others) in order to say something about the human but in the process risk overlooking and disregarding their uniqueness as other. In other words, our "negative philosophical anthropologies" always need others who are then marginalized in terms of their own alterity. In an effort to respond to this problem, Coeckelbergh turns to Levinas, but he does so recognizing that this turn is not without its own problems and complications insofar as Levinas himself would most certainly have resisted such a development. For Coeckelbergh, however, the important question is not whether Levinas would have endorsed this application of his philosophy or not. What needs to be asked about is much more fundamental: How does the other appear? How is a concrete encounter with the other—any other . . . human, animal, machine, or otherwise—possible in the first place? This fundamental phenomenological inquiry leads Coeckelbergh to propose an ethics of unconditional hospitality. Such an ethic, which is similar to Marcondes Filho's proposal but which also challenges its limitations, does not make prior distinctions between who or what can or should be other. It is an alternate mode of comportment toward alterity that is, in a kind of reactualization of Romanticism, open to other kinds of others in such a way that respects their otherness and that does not domesticate the other by immediately naming or assigning it a place in our ontological schemes. To return to Heidegger, the moral innovation that is proposed by Coeckelbergh would be a posthumanist ethics of *Gelassenheit* or "letting be."

If Coeckelbergh introduces a posthuman ethic of radical hospitality that is open to the other and to other forms of otherness, like the animal, the machine, and the alien; the final chapter by David J. Gunkel uses this insight to turn the question of alterity on itself. Gunkel's proposal for "thinking otherwise"—a phrase that should be read as indicating a different mode of thinking that is able to accommodate other forms of otherness—shares Coeckelbergh's efforts to open up ethics, in general, and Levinasian thought, in particular, to different modes of alterity. Despite the warnings and arguments provided by many of the other contributors to this volume, we seem ready and willing to consider technological objects to be other—not just a kind of surrogate pet (like Sony's robotic dog, AIBO), but also a close personal friend, confidant, and even paramour. In doing so, we are apparently in danger of substituting the technological interface for the genuine face-to-face encounters we used to have (and should continue to be having) with other human beings. Gunkel, however, flips the script. The problem with our socially situated and increasingly interactive devices, he argues, is not that they substitute a machine interface for the face of the Other. Instead, it is in the face of these others (e.g., animals, social robots, non-player characters in computer games, avatars, algorithms, etc.) that we are challenged to reexamine critically who or what is, can be, or should be the Other. In other words, if the intimacy of network connections and socially interactive mechanisms appear to threaten human sociality and communication, it is it is not simply because these devices are being substituted for real human contact, but because it is in facing the question of the alterity of the machine that we are forced to face-up to and reconsider the violent exclusions and defacings that have always been made in the face of others and in the name of ethics. Consequently, instead of formulating another "liberation movement" (Peter Singer's term) and arguing for the inclusion of previously excluded others in our moral calculations, Gunkel suggests that what we see in our encounter with the face of the machine is nothing less than a fundamental critique of and challenge to ethics itself.

4. ALTERNATE ENDINGS

In the end, this is most assuredly a philosophical book. As such, it adheres to the characterization provided by Daniel Dennett and quoted above. The book, therefore, will not provide readers with a set of prepackaged, sound-bite solutions to what are assumed to be the important questions concerning the fate of *the face of the other* in an age of increasingly sophisticated technological affordances and configurations. Instead the book seeks to identify and articulate the questions that need to be asked in the face of the other and in the face of other configurations of otherness. These questions can, at the end of the introduction—which like the introductions written by G. W. F. Hegel (1977) are always after-effects presented in advance of what follows, or what Slavoj Žižek (2008, 209) characterizes as "retroactively (presup) posited"—be collected into three categories, demarcated by the three keywords that comprise the book's subtitle:

Communication—What is communication? Or better: *Was heißt Communication*? What is called or what calls for communication? According to James Carey (1992, 15), what we call "communication" can be understood from two different perspectives: a transmission view and a ritual view. The transmission view concerns the conveyance of messages from senders to receivers. It is, Carey argues, the most common view of communication in the Western philosophical tradition. It is the conceptualization that is modeled and operationalized by Claude Shannon and Warren Weaver in *The Mathematical Theory of Communication*. And it focuses attention on matters that are quantifiable and that can be made computable. The ritual viewpoint, Carey (1992, 18) argues, looks at communication otherwise. It draws on and calls attention to the common root of the words "communication," "community," "communion," and "commonality." It, therefore, focuses attention on a different set of issues, problems, and possible solutions; looking not at quantifiable data but at the different qualities of experience that are formed when we are exposed to and come into contact with one an-other. The essays contained in this volume do not seek to resolve the tension between these two views of communication, but exploit the differences between them in order to formulate more sophisticated understandings of what is called and what calls for communication.

Technology—Do recent advances in technology deface the Other through the imposition of the technical interface or the seemingly impenetrable surface of the screen? Or does technology make available other facets of otherness that challenge our "human, all too human" assumptions about who or what can or should be other? Again the ten essays collected here provide a range of different versions of and approaches to these questions. Some try to shelter the Other from the incursion of technological mediations that threaten to limit, impede, or even foreclose the unique experience of alterity. Others argue that it is in the face of others, especially in the face of increasingly capable and socially interactive technologies, that we confront a fundamental challenge with new and unique forms of otherness. Still others, stake-out a position in between these two extremes, parsing the question of otherness in other, perhaps more nuanced ways. In addition to these different approaches, there is also one other question in play. In the face of recent technological innovation, is the proper response to retreat from technology through a kind of critical asceticism that responds to the opportunities and challenges of these affordances by just saying "no" to applications, like Facebook and Instagram, and social robots, like Siri and Jibo? Or is the answer found in and by technology itself? In other words, where can we find, what Heidegger (1977, 28) called, "the saving power?" Is it located outside and beyond the realm of the "danger" that is presented by technology? Or is it to be found in the medium of the danger itself?

Other Subjects—Is "alterity" a noun? Or is it a verb? Is "face" something that is possessed by the Other? Is "alterity" a property? Or is "alterity" an event that happens in between me and an-other? Is "face" less a property of the Other and more a facet of the event of an experience of otherness? The essays that comprise this volume take very different positions with regards to this question. For some, alterity pulls in the direction of a noun. It is the substantial quality of the Other that makes the Other other. And support for this position can be certainly be found in the work of Levinas, who often deploys the word "face" as if it were a kind of personal property. Other chapters insist on the verbal aspect of the word "alterity," calling attention not to the subject of the Other but to the event of otherness. We might say, therefore, that this difference—noun or verb—concerns the very being of alterity. Is alterity a kind of persistent substance or property? Or is it a kind of temporary *being in time* that is always open to various occurrences and becomings? This can perhaps be best articulated by calling upon and leveraging the ontological nuance available in the two Portuguese verbs for the single English "to be" and the German "sein"—*ser* and *estar*. The question of alterity, therefore, can and perhaps should be written twice: *O que é a alteridade?* [What is alterity?] and *Como está a alteridade?* [How is alterity?].

REFERENCES

Benso, Silvia. 2000. *The Face of Things: A Different Side of Ethics*. Albany, NY: SUNY Press.

Carey, James. 1992. *Communication as Culture: Essays on Media and Society*. New York, NY: Routledge.

Dennett, Daniel C. 1996. *Kinds of Minds: Toward an Understanding of Consciousness*. New York, NY: Basic Books.

Hegel, Georg Wilhelm Friedrich. 1977. *Phenomenology of Spirit*, trans. A. V. Miller. Oxford: Oxford University Press.

Heidegger, Martin. 1962. *Being and Time,* trans. John Macquarrie and Edward Robinson. New York, NY: Harper & Row.

Heidegger, Martin. 1977. *The Question Concerning Technology and Other Essays,* trans William Lovitt. New York, NY: Harper & Row.

Manovich, Lev. 1995. What is Digital Cinema? http://manovich.net/content/04-projects/009-what-is-digital-cinema/07_article_1995.pdf

Moore, George E. 2005. *Principia Ethica.* New York, NY: Barnes & Noble Books.

Shannon, Claude E., and Warren Weaver. 1963. *The Mathematical Theory of Communication.* Urbana, IL: University of Illinois Press.

Žižek, Slavoj. 2006. Philosophy, the "Unknown Knowns," and the Public Use of Reason. *Topoi* 25 (1–2): 137–142.

Žižek, Slavoj. 2008. *For They Know Not What They Do: Enjoyment as a Political Factor.* London: Verso.

Part I

THE FACE OF THE OTHER

Chapter 1

Countenance—Mask—Avatar

The "Face" and the Technical Artifact

Dieter Mersch

1. A QUESTION OF RELATION

What does it mean to "meet" a robot? What happens when we contemplate it, look into its face or stare into its "eyes?" How do we read the face of an avatar or understand its expression; in short, how do we respond to it? Does gaze meet gaze? What are we reacting to when we dialogue, solve problems, or fight with it? It seems that the face—German, *Antlitz*, also countenance or *visage*—is neither an image that can be animated at will nor a simulative screen open to mimetic play, but rather the opening to the Other, an "abyss" or "alterity." Emmanuel Levinas (1969, 197) described the face as the "first revelation of the Other" to whose expression we "respond." From the very first moment we communicate with one another, we throw ourselves into a veritable labyrinth. We become flustered or lose ground the instant we make contact. We carefully get a "feel" for each other, even if we only catch a sideways glimpse of one another in passing. Continuously, the face creates confusion; it takes hold of us and echoes within us long after we have shifted our attention elsewhere. To meet another person is thus, as Levinas so aptly put it, to be "kept awake by an enigma" (1998, 111)—open and ready to be unnerved or, to the contrary, scared-off or repulsed.

The experience of the face is met by a fundamental "inindifference," as Levinas said. The double negation emphasizes the impossibility of disinterest. For this reason the face is always like an interruption, a "trap" or an attraction that leaves an indelible impression upon us and reminds us that we are first and foremost *social* beings and dependent upon others. We refer to others, desire them, and share a "world" with them, whether we want to or not. The "nakedness" of the face, again Levinas's (1969, 75) formulation, the fact that we usually present it unprotected—disclosing its insufficiency and distress, its inherent vulnerability because it exposes the bareness of our existence—also implies that it "concerns" us *(regarder),* attracts the gaze and expects respect (égard), but at the same time demanding distance and restraint. Levinas speaks in this context of "supplication," the expression of a "first word, 'you shall not commit murder.'" It is almost impractical to destroy a face. For this reason, confronting the countenance also has an ethical dimension. Even if we are oblivious to this fact, every face brings puts us on the trail of the initial experience of sociality that tells us, in principle, you are like me.

Is the above also true for our interaction with robots or avatars? Can they similarly become a "counterpart" that we meet "face to face," like the intermediary incarnation of Indian mythology from which the motif of the avatar stems? Or is there a fundamental separation, an unbridgeable gap, as if two inaccessible territories faced one another? Masahiro Mori (2012) made a similar claim as early as 1970, a claim that reappears in Jasia Reichardt's 1978 study *Robots: Fact, Fiction* and *Prediction.* Mori theorized an "uncanny valley," an unease that erupts when technical artifacts become too close to us and their appearance all too familiar. Our acceptance or approval of robots—and avatars—correlates directly, Mori postulated, with their lack of resemblance to us. The more they look like humans, the more they awaken revulsion. It seems we can accept animated dolls, animals, or automatons only if they are not "ghosts" or doppelgängers that we are unable to control. Otherwise the question of their autonomy arises, their "social" status and the respect due to them, depending on the specificity of their differences. The "uncanny valley" is our discomfort at their sameness and it exists in principle for all technical or digital devices with which we interact. Thereby the true question is what exactly does "inter-action" mean, in particular what is "in between" and what can we "share" in its *spatium.*

In the same vein, the relation between human and machine or simulation and "life," or the possibility of their mutual confusion, is also up for discussion, in particular the question of photorealistic rendering or 3-D models of "people." The problem is not so much whether machines—or computers or robots—can think. Alan Turing attempted to determine just that with his test, which tellingly works with curtains behind which the concealed competitors are asked to make decisions meant to reveal which one of them is, without a doubt, a technical structure (Turing 1950; see also Hayles 1999, xi ff.). Decisions however take place below the threshold of the discernible, for which reason Turing was content to contest that once a machine passed the test, we had no more reason to deny that it "thinks." However the entire construction already presupposes a decision-logical arrangement and thus encircles its own argumentation. The error is thus rooted in the set-up of the experiment itself, which operates within a binary logic where undecidability implies undifference—but undecidability and indistinguishability are not the same (Mersch 2013). Indeed the greater problem by far is situated *before* thought at the level of perception, which Turing intentionally precluded from his experiments and which reveals, in a reflection on artificial skin or a "dead" eye, a difference that thwarts deception.[1] If we speak only of "similarities" we have already accepted that distance and distinction. But the question is whether—perhaps in the near future—analogs will exist that not only defy understanding and recognition of distinctions, but also exhibit reactions and affects similar to our own, so that, as some science fiction movies suggest, we despair of trying to detect them. Or put another way, might we, in our meetings with artifacts—avatars or acting and speaking machines—possibly develop the same deep-seated desire and disquiet, and repulsion and attraction we experience in confrontation with human "people?" If we answer in the affirmative, we need moral standards for these meetings, a second posthuman ethics to stand alongside ethics as the *prima philosophia* that Levinas attempted to formulate. In developing robots and avatars are we confronting ourselves with digital twins that, once and for all, deprive us of our uniqueness—a further metaphysical humiliation beyond Copernicus's revolution, Charles Darwin's model of evolution, and Sigmund Freud's theory of unconscious—and reawaken doubts about our singularity? Or is the human face in contrast unique and irreplaceable—enjoying a distinct form of encounter that could never occur with technical devices, images, or other objects and apparatuses?

2. AURA AND ALTERITY

Today, there seems to be a tendency to believe that the former is at least likely. Posthumanist ethics, which aim to oust human beings from their central position—either as cognitive subjects with privileged access to the "world" and to the "truth" or as actors in a reality populated in the main by, as Graham Harman (2007) put it, Other objects—seemingly postulates a symmetry that denies both human exceptionality and the genuine asymmetry of human sociality. In that paradigm, there is no reason to favor humans over artifacts or things. Important instead is the analysis of mutual networks and relations or, in the words of Timothy Morton (2008, 2010), of unnatural ecologies, in which people are at best one node among others. As the problem is too complex for the scope of a single chapter, the following focuses on one key aspect, the problem of identity or difference in the relationship between "alien" objects such as robots or avatars and human beings. By "robots and avatars," I mean machines that seem to act and communicate autonomously as well as digital figurations that serve the purpose of resembling humans or taking over some of their features and functions. This analysis is further restricted to a concentration on, paradigmatically, the similarity or dissimilarity of the human face and the "face" of the avatar, whereby the mask shall serve as both parallel and mediator. The goal of these explorations is to disturb the seeming plausibility of a posthumanist movement that posits itself as avant-garde and assumes their sameness. In contrast, the chapter at hand insists there is a difference in our manner of relating—if not necessarily in our relationships.[2] It claims that the experience of the face is a "model" for an experience of alterity that *precedes* all experiences of things or artifacts, even when they look deceptively like us. In short, I uphold the primacy of an asymmetry that stems not from the ontological supremacy of humans over things, but from our relationships to others, which are different from our relationships to objects or non-humans, even when the latter are artificial systems which looks like humans. In fact, we owe our relationships to things or automatons first and foremost to our primary relationship to the Other, so that a constitutive difference opens between the Other and the others (in the sense of the social). Or, put another way, we are interested first in others and only secondarily and derived from there in the world we share with them or in a "nature" that presents itself to us as that which we did not make and are nevertheless a part of. There are, therefore, two different modes of relationship, the relation between humans and non-humans (things), which is intentional, and the (social) relationships between humans, which are either active or passive but both founded in their prior ability of responsivity. It is for this reason that myths personalized natural things, in order to move them closer to our understandings and behaviors.

A prime example of this distinction can be made by examining the meaning of the face. For us, as literary scholar Peter von Matt (1983) has said, the face is the "condensed image of the *humanum*" and at the same time the absolute heteronomy that embodies that which is completely inscrutable. In the face we meet the trace of the "God who passed" as Levinas (1986, 359) pointedly stated, for which reason its "nudity" and "infiniteness" demand an ethical stance. It is the prerequisite and the basis for every social relationship. Walter Benjamin (2010, 19) claimed almost the same, calling the "human countenance" the "last entrenchment" against the disappearance of the "aura" in a world of "technological reproducibility." The face's "fleeting expression . . . beckons from early photographs for the last time." "It is" Benjamin continued "no accident that the portrait"—like painting in early modernity—"is central to early photography," rather than the perfect illusion of a space constructed by means of mathematical perspective. Similarly, Béla Balázs (2010, 38–41)—as later developed by Gilles Deleuze in his books on cinema—believed the facial close-up to have a particular "emotional" status and claimed it was more central than landscapes. He even believed facial expression to be a universal language of emotion that could not be replaced by any aesthetic of the sublime, no matter how impressive.

3. RESPONSIVENESS AND THE PRIMACY OF ANSWERING

The difficult concept of the aura—using Benjamin's (2010, 19) definition of "the unique appearance of a distance, however near it may be"[3]—thus coincides with the way in which alterities, as the site where all types of relations are constituted, are at the center of Levinas's critical phenomenology. Levinas sought in particular to bypass Husserl's insistence on an always subjective "intentionality" by means of privileging answering or the more passive "responsiveness" (Waldenfels 2007).[4]

Here *responsio* comes before *intentio*, thus intentionality is rooted in responsivity. Relating is therefore not an *actio* of an acting subject, but develops from the site of the Other as a primary event of answering—without us knowing what we are responding to. The same is true of the aura. It cannot be forced, Benjamin (1997, 147ff.) states in his analysis of Charles Baudelaire, but appears as a recusant surprise. Likewise, comprehension of a face is not a conscious act, as long as I turn toward it curiously. Rather its attraction *happens*, it takes me by surprise and pierces me with the rapidity of an arrow. For this reason the face—like the aura—cannot become the *object* of a gaze, just as the idea thereof cannot come out of myself. Rather, it never truly arrives, remaining, as Levinas also said, "infinitely" transcendent. This infinitude and transcendence is its most radical denial. That experience, drawn from twentieth-century Jewish scholarship, can by no means be dismissed as exceptional. To the contrary, its "anticipation" of the "absolute otherness" of the Other gives it clear primacy over every other form of relationship, especially encounters with things, technical objects, or avatars, which can at best be deduced from it. To be more exact, it is important to distinguish between relationships and relations. The latter can be formalized by functions, they prove to be *Zu-Ordnungen* (attributions, literally toward-orders), correlations or configurations of points that express affiliations, not *Zu-Wendungen* (attentions, literally toward-turnings) around which our relatings have always gravitated.

The same perspective is incidentally shared by other philosophical positions that see the experience of the face as the primary site of human relationships. As only one example, allow me to cite Georges Bataille (2002, 63): "Nothing is human in the unintelligible universe outside of naked faces which are the only open windows in a chaos of strange or hostile appearances. Man only escapes his insupportable solitude at the moment when the face of one of his fellow men emerges from the void of all the rest." Art historian Hans Belting (2013, 25ff.) has made similar observations about faces and masks. The critical point being—and this is what makes these analyses so fruitful for our context—that the relationship between symmetry and asymmetry is turned on its head. It is not we who, through the power of our decisions or desires, turn our gaze to the Other; we are already in their horizon and have entered their realm *before they arrive*, so that every relationship presumes the "originality of the face" without which, Levinas pointedly stated in *Totality and Infinity* (1969, 202) language "could not commence."

That also means that there is no dialogue or communication worthy of the name that was not first "attuned" by the primary experience of the Other and their face. This is why the response precedes the *intentio* and gives intention its specific modality. The face becomes a site of non-intentional "immediacy," it is "evidence that makes evidence possible" as Levinas (1969, 204) continued, and is not transferrable to objects, because the face is neither a medium nor does it speak to us through a "mask." The face provokes us, it comes at us and holds us captive in the literal meaning of *fascinans*, and solicits a reaction or positioning that first makes us to that which we *are*.

4. IMAGE AND FACE

This structure of primordial non-intentionality, which I would like to propose as the third initial thesis of this exploration, can also be applied—as seen in Balázs and Benjamin—to pictures, photography, and film, as long as they show faces. The "originality" of the face and the concurrent initial responsiveness seems to be etched into these depictions from the very beginning. Observing or watching means participating, in the meaning of both taking part and having recourse to. Nevertheless it seems logical to assume that in our era of digital plasticity, with almost limitless possibilities for manipulating or editing images with programs such as Photoshop or through mathematically generated random simulations, an effect takes place similar to that which Benjamin diagnosed as the "decay" or "destruction of the aura" in the age of "technological reproducibility." Are we still dealing with faces, with a "countenance" in the literal sense of a counter-gazer? In Benjamin's diagnosis, reproduction causes objects, on the one hand, to lose their "unique existence," while, on the other hand, the experience of singularity is subject to a ruthless appropriation or "getting a hold of" objects in a "facsimile," (*Abbild*) which he claims "differs unmistakably from the image (*Bild*)" (Benjamin 2010, 15–16). Jacques Derrida has made similar assertions about technological repeatability, which can only ever produce the same—the series and thus the stereotype. Apparently, image and facsimile—the image of apparition and the facsimile of representation—or that which Husserl named the *Bildobjekt* (image object), need to be separated from one another, whereby the loss of aura marks the defamiliarization of all relationships, which vanish when an object loses its uniqueness through technological reproduction. The generation of virtual images in the form of avatars and the destruction of a person's aura go hand in hand because the transformation of the face into an image and its free alteration by means of digital figuration leaves out that which Levinas (1969, 192–193) discussed as "alterity"—the moment of radical "transcendence" from which stems the always ethically charged rigorousness of having to respond. We can also express this as follows: That which paradigmatically resounds or "resights" within us from the human face and is perhaps, as we shall see, still present in the enigmatic form of the mask—the alterity of the face—denotes exactly that mysterious dimension that fades in confrontation with the technical apparatus or avatar or robot and that haunts us with increasing unease. Our *relationship with* the face differentiates it from the avatar. We act *differently* with people than with artifacts and machines. We relate to them in *another way*. We see them differently, consider them differently, and handle them *differently*. We must therefore assume their genuine incomparability. We are, thus, confronted with disparate relational modalities.

There are two branches of philosophy that intersect at the site of the face in the as yet tentative analogy between aura and alterity that guides this exploration—aesthetics and ethics. For this reason, all associations with ontology should be rejected, in particular with posthuman "object-oriented" ontologies and relational "ecologies" as formulated by Bruno Latour's successors, in particular by Timothy Morton, Graham Harman, Levi Bryant, and Ian Bogost. The relationship to the face, the spark of all possible relations, is radically different from these theories. It is not ontological, but both sensuous and transcendent, and thus from its beginning aesthetico-ethical. Yet "alterity" remains a concept that is difficult to delineate. Not even truly a concept, it evades comprehensive definition. Its chronic under-determination stems from its duplicity, located as it is between "draw" and "withdraw" (*Zug* and *Entzug*). Alterity, following Levinas, cannot be seen "as" something, because this subsuming circumscribes and cancels out the difference of the Other. It cannot be traced back to a sequence of signs and thus defies the logical order and, with it, predication, the listing of distinct characteristics or the ascription of meaning. To delineate in this case would be to possess. But the face cannot be the *object* of a gaze or of an understanding, just as little can I have the "idea" of the face within myself. It cannot be made just by imagination. Its withdrawal thus, on the one hand, points toward pervasive negativity. Any positive characterization, any address, should I attempt one, would necessarily get caught up in contradictions, because the moment of appropriation erases that which it attempts to grasp. The "absolute other" that can be named is no longer an Other, Levinas reminds us. For we would then own it through the category or the statement, having subjected it to the power of meaning and defamiliarized its otherness. Is this not also true for the avatar and the robot—to the extent that we have given them characteristics and chosen them as figures of projection? Even when they present themselves as autonomous actors, they are always only others that have been trained *by programs*—their seeming alterity is a mathematically defined, if randomized, endless chain of attributes. In principle, they remain calculable. On the other hand, and at the same time, it proves impossible to deny the experience of alterity. It cannot be negated, because it first provides and therefore precedes any possibility of relationship—as long as that which we call perception or sociality ensues from its primary responsiveness. It "gives" a "gift" that can only be accepted as such.[5] For this reason we can speak of a genuine duplicity of "drawing" and "withdrawal," and of "distance" and "affection." This is reiterated in the simultaneity of alienation and desire. Its dynamics, as regards artifacts and other technical things, takes the form of a claim of usefulness. That is to say, it amounts to nothing more than *function*.

5. FROM *ACTIO* TO *PASSIO*

This also means that the beginning of thought, of intentionality, and also of our being-in-the-world lies literally *elsewhere*. It is founded in something that cannot be adequately thought or described. Nor does it exhibit distinctive characteristics, but meets us as a trace, with all the concurrent indirectness of the term. It can neither be identified as "certain things" or "objects," as Harman suggests, nor can it be rejected or ignored. Thus it remains a double negative. We understand this doubled negativity as the display of a fundamental *Unter-Schied* (for Heidegger "difference," literally, under-separate) that sets the conditions for the identity of "is" and from which comes our access to the world, to things, and to other human beings. Giving precedence to alterity means prioritizing an *Unter-Schied* or assuming a *différance* that opens the possibility of relationships and thus also "accessibility." And while relational ontologies such as the actor-network theory assume the primacy of relations that bring forth *relata*, they do not cross this threshold, because they do not take account of the modality or plurality of varying forms of relationships, or of that which first opens them. Instead, relationships seem to preexist; there is no question of what constituted them. An undifferentiated network of relations is supposed as that which can only be adequately described in a formal mathematical manner. I instead am attempting to bring that which opens up, and has made these relationships possible, into play. In this way I hope to ascertain *what* is responsible for the differences in relations and where they come from. For this reason I insist upon the difference that arises from the disparity of our relationships to things and artifacts as opposed to people. As Levinas (1969, 202) emphasized, the "face to face also cuts across every relation," it causes a change in reference or a "relational turn"[6] that changes direction in face of the Other and moves from *actio* to *passio*. Finally, it touches us, and from the very beginning adds an ethical dimension to our relationships as well as to the possible meaning of our relations with artificial objects or avatars.

To better understand this experience and its consequences we shall continue to explore the appearance of the face, which for Levinas is the nucleus of that which the otherness of the Other—between lucidity and opacity—*presents.* It is a presentation of which he says both that it cannot be experienced without the senses and that it never arrives. We are dealing with transcendence *within* the sensuous—a crossing or opening that has no correlate in reality. "It can be stated neither in terms of contemplation nor in terms of practice," Levinas says of the face in *Totality and Infinity* (1969, 193). "The relation with the Other alone introduces a dimension of transcendence and leads us to a relation totally different from experience in the sensible sense of the term." This also stems from the fact that the face acts as a model for the *first* relationship into which we are literally born. Before objects and their disparity, before any apprehension of similarity (the condition for comparison), the face is that which turns toward us and through which we learn how relationships are formed—underscoring the passivity of our existence. The face is revealed to us, it speaks to us and names the site of our primary passion. From the face we can trace the first facets of our dealings with those closest to us and also with things and technological devices. Robots and avatars, I therefore claim, only have "faces" for us, if we want to call them such, because we have always related to faces—the faces of others and their enigma. We thus have a preexisting experience of "faceness" that we call upon and that allows us to identify artificial faces as such.

In German, it is possible to mark the identity of this "facey" passive face by distinguishing between *Gesicht* (face) and *Antlitz* (countenance), a difference that does not exist so clearly in other languages and in particular not in Levinas, who speaks always of *visage*. "Face" denotes the sensuous apparition, while "countenance" marks the Other that shine through it, the transcendence of alterity that *presents itself*. "What we call the face is precisely this exceptional presentation of self by self" (Levinas 1969, 202). This self-presentation is the "trace" or "aura" that transcends every face that we are able to experience concretely, whether it touches us only in passing or demands a response. Languages have found an extraordinary number of different words for the phenomenon of "faceness." In English, alongside the main term "face," there is "visage" and "countenance" as well as pejorative terms such as "muzzle," "mug," or "kisser" and terms related to expressions such as "grimace." In French, alongside "visage" there is "face" or the frontal surface thereof, the diminutives "minois" and "frimousse" as well as disparaging terms such as "bouille," "binette," or "gueule." These terminological variations mirror the micro-facets of the face and its varying apparitions. Despite their lack of respect and occasional rejection, it underlines the importance of drawing facial diversity in language. It is impossible to ignore that these terms present the nuances of human beings and their faciality. The terms are situated in the visual play with image and figuration, which again conceals the irrepressible countering of the countenance, which remains in its variety transcendent.

The changing use of the varying terms makes visible experiences and the sedimentations of history, which we connect with ambivalent meetings with the faces of "friend" and "foe," linked to "attraction" and "repulsion" as well as to "trust" and "alienation." At the same time, *visio* is dominant or seeing in the sense of recognizing. Clearly the face and its varying forms interest us greatly and become a paradigm for our, negative or positive, relationship to the world—unlike objects, about which we are usually indifferent.

6. THE FACELESSNESS OF THINGS

This holds especially true in observations about the German etymology of the two central terms *Gesicht* and *Antlitz*. Their relation to one another mirrors the difference between our relationships to artifacts and to people. *Gesicht* originally denoted a person's identifiable face and is linked to a long history of identity and identification reaching back to early modernity and also resulting in legal consequences. While in Middle High German *Gesicht* still meant the "vision" of second sight (in German *zweites Gesicht*), since the sixteenth century at the latest, it has been linked to questions of the constitution of the individual and the identity of the person. It became literally *ge-sicht*, that which has been sighted and thus can be recognized and identified. From the beginning, then, *Gesicht* was connected to an image or profile that allowed for repeated recognition. The older term *Antlitz*, by contrast, was kept only in highbrow language and poetry. The Old High German *ante-lizzi* refers to a counter-gazer, with its prefix *ant* "against" or "toward" and *lizzi* (Middle High German *litze*), a verb that has been lost in German with the exception of the dialectical *lugen*, "to see." The *Antlitz* is therefore the gaze that comes toward us and demands a response. Behind this difference hides a complete philosophy: the latter is associated with that which comes to us and which we can meet only passively, while the former denotes a conscious "addressing of" that domesticates "the seen" through the gaze. In the *Antlitz*, it is the gaze of the Other that positions and animates us, while it is we who look curiously at the *Gesicht* and put it, literally, under observation. From this time on, the *Gesicht* has been linked to the idea of the "person," which once was, in terms of the Latin *persona*, a mask, while the *Antlitz* had a transcendent meaning comprising both the Other and the numinous. Thus *Gesicht* and *Antlitz* allow us to differentiate between two dimensions of faceness, identity and alterity. And while *Gesicht* emphasizes that which is looked at in order to make it into the object of a gazing subject, the *Antlitz* comes from the Other, looks at us, and robs us of our sovereignty. This observation can be directly applied to artifacts and technical objects which, like puppets or avatars, have at most "facial images" or *superficies,* surfaces made of flattened and derivative faces (*Gesicht*) without a countenance (*Antlitz*). That is the source of the disparity, or fundamental asymmetry, of the relationship between humans and things.

Moreover, if we look at the ancient languages we find a further field of associations. The Latin *facies* denotes not only the face but also the apparition or the appearance with clearly passive connotations—it names the moment that is given us and is related to *facetus*, the elegant or radiant, in which we again find the experience of the aura. The ancient Greek analog is *prosōpon*, within which is *op,* the gazing "eye."[7] *Pros-ōpon* is then that which is in front of the eye, turned toward us, also making a person's status visible. This is found again in terms such as *opōpē* and *enōpē*, "face" and "view." Interestingly, *prosōpon* also denotes the theatrical mask that not only hides the face of the divine, but also embodies the stylized expression thereof.[8] Thus, the face *is* the mask with all of its paradoxical showing and hiding (Weihe 2004, 35–36). Hence, "face" manifests not only the visible or accessible, but at the same time the invisible and inaccessible that hides a secret. The mask is consecrated because in ancient theater it is always both the image of the god, and thus the face *par excellence* that acts within the person, as well as the image of the person ruled by the caprice of the gods.

Here too we can see the duplicity of *Gesicht* and *Antlitz*. For the mask both conceals an alterity and presents an expression, the "façade" of the face, *for* others. And as in Roman theater *prosōpon* was translated as *persona,* in which the sound of the voice (*sonor*) speaks through (*per*) the mask and must be distinguished from *facies*, the actual face. Christian iconography defines the *persona* of Jesus Christ, the Holy Face that in becoming human wears the mask of God, as conversely it stands in for all mortals (Belting 2013, 47). And in fact, in the oldest European traditions, the duplicity of face and image has always dominated, as Hans Belting (2013, 26) has noted: "Our faces awaken and become images as soon as we gaze and speak. We stage ourselves with our faces . . . We communicate and represent ourselves."

At the same time, we cannot completely control these images our face is part of us, it reveals our feelings and our character, it *presents* us. We try in vain to form these, but something—we know not *what*—evades our control. It is that which we have termed *Antlitz*, the etymology of which suggests that it cannot be an artifact, nor can it be replaced by an artifact, even if we wear false faces in public and put on societally expected expressions of approval or rejection. This fact has been useful to both classical physiognomy as well as to modern semiotics of expression. Belting (2013, 120) has diagnosed the social as the site where face and mask come together, for "only the mask [can] in the long term represent a face, while a face never rests and therefore cannot be pinned down." And yet, I would like to emphasize, something fundamental is missing here, for no matter how mask-like a face may be, it still carries the "authenticity of the face" of which Levinas (1969, 202) has said it "circumvent[s] the ambiguity of the true and the false." It conceals itself beyond the seen in the unseen. One could call it that which remains while the face changes. It resembles a hollow, a negativity that like a mirror reflects that which makes a face to this particular face. It reveals itself when we lose our countenance or lose control over our expressions. Such phenomena are unknown in artificial faces. Instead, there is nothing "behind" them. Unashamedly and mysteriously they reveal themselves as technical designs and our gaze loses itself upon them without penetrating or traversing them. Avatars and other 3-D or computer animated faces thus perform a tyranny of absolute visibility, they iconize the avatarian "facial image" and transform it into the image of an image. For this reason we do not gaze upon it as much as our gaze rebounds off of it and scatters on the surface of its pure planarity.

7. MASK, AVATAR AND ROBOT

A series of further questions are linked to the relation between mask and face as regards their artificial counterpart, the facial image of robots and avatars. Their difference provides a framework within which we can discuss the much trickier question about the relation of a person's "image" in portraits and the digital production thereof. Robots and avatars may sometimes be described, using the metaphor of the masquerade, as the "great unknown" or as "mediators" between the real and virtual worlds—the incarnation of Vishnu in Indian mythology (Wesserly 1997, 177f.)—but masks are literally in another category. For this reason we speak of *superficies* or facial images as surfaces in contrast to masks, because we are not dealing with *mimesis*, but with calculation or construction according to geometric parameters. Whether an avatar is a drawing (vector graphic) or a space curve (3-D graphics), it is always a mathematical figuration. We are thus confronted with *three* terms and *three* differences that generate varying semantics, face, mask, and avatar/robot and the alternating reciprocal differences between countenance and face, face and mask, and mask and avatar/robot. All three prove incommensurable with one another. For while the countenance (*Antlitz*) shines through the face and is thus also perceived, confronting us with a genuine duplicity, face (*Gesicht*) and mask interpret one another, because masks do not conceal faces as much as reveal "another" or "second" face that directly invokes the use of vision in the Middle Ages. The mask disturbs the experience of the social, which the face opens only to ban it into the night of death and animality. The ancestors, the ghosts of animals or of supernatural powers speak through masks and, according to Belting (2013, 46), the dead are dependent upon masks after losing their faces. Masks therefore contain a genuinely uncanny element, which Freud diagnosed as the canny that must be repressed and takes on a life of its own. Thus sociality, animality, and death are irrevocably linked. The mask reminds us of this connection. For this reason it is, says Bataille (2002, 64), "present before me as a likeness, a fellow man, and this likeness, which stares at me, has taken into itself the figure of my own death."

On the other hand, mask and avatar seem to resemble one another, since both have the character of an image. The mask, as Belting (2013, 44ff.) repeatedly emphasizes, has its roots in the effigies of ancestor worship and avatars have also been compared to marionettes or mannequins and represent the player in computer games (Beil 2012, 11ff.; Calvillo-Gámes and Cairns 2008). Nevertheless they are no more than a moving image within a program that mathematically creates "lively" and expressive movements. We are dealing with an algorithmic structure that turns the avatar into a mathematical object. In so doing, it links statistical parameters with morphing, the practice of digital facial rendering[9] paired with the facial action coding system, which creates emotive expressions. Unlike the mask, the avatar is the result of a complex description into which has gone not only techniques of abstract visualization, graph theory, and geometrical projection, but also knowledge from statistical psychology and the semiotics of emotion. The facial likeness is the product of a rule or diagrammatic form. As a digital schematics, it is founded completely in mathematical encoding and its surface adheres only to an algorithmic/semiotic regime that, I hypothesize, contains nothing more of the mask's manifest symbolism, ritual, and religious practice, but instead obeys calculable actions. While the mask conceals the person wearing it and is, thus, a *medium*, robots and avatars refer to no one, they operate blindly for themselves (Mersch 2016).[10] In particular the construction of avatars, founded as they are in computer games, is based on a series of characteristics or movements that can be chosen; they are a series of choices that nowhere attempt to deny their formalism. From the beginning, their algorithmic structure therefore fits into a model of logical decisions and so remains—based on decisions *we* make—in the narcissistic or "egologic" space of our projections. The mask in contrast exhibits manifest social and therapeutic functions. By being death and animal, it confronts us by proxy with the Other of our self, with that which we neither possess nor control—our "nature."[11]

This last insight puts us on another track. As regards their iconicity, the intersections of the face, the mask, and the facial images of avatars and robots are obvious. But an experience of alterity is key to the face and the mask that is lacking in (the design of) technologically generated artifacts. Just as the face, despite the social encoding of mimicry and expression, nevertheless retains and reveals the singularity of a countenance (*Antlitz*), the mask too—which can never be completely opaque because it "lives" to a great extent from the gaps at the eyes and mouth—is animated by the foreign gaze and foreign voice. Every mask needs a wearer who first allows it to unfold its particular magic and "come to life." In this sense, the mask is a "seeing through the eyes of the Other" and a "speaking through the voice of another," while nothing seems surprising in an avatar no matter how lacking in motivation and how unpredictably[12] it behaves. Behind the mask is always the shadow of an otherness—which makes it spectral—whereas nothing but a computer program "gazes" or "acts" from the back of the avatar or robot (without any true gaze or acting). The puppet must still be led by the hand of a person; this no longer holds for automatons and avatars. This means that, no matter what one understands by "mask"—and its theories are as diverse as ethnological approaches (Caillois 2001, 87ff.)—it is always experienced as a "haunting" by an alterity to which it lends alienation. A "spectrality" gazes from the mask in which the visible and the invisible meet, turning them into something indeterminate. Technical objects like automatons and programs in contrast manifest only the impenetrability of the technical, in which we at best meet only ourselves. For this reason, the *superficies* of avatars and robots describe only an immanence (Han 2015, 10) whose signs circulate on the surface, while the mask, no matter how much it remains symbolic and includes the figure, lives from the configuration of the Other—a direct contrast to the mathematically modeled figure. No robot can respond to us or have a relationship with us that extends beyond its abilities, just as little as avatars can reveal their "secrets" to us. Rather "avatarization" turns every "face" into a display of what Benjamin (2010, 19), referring to the unique presence of the aura, referred to as an object's "exhibition value." It becomes a tableau, an empty slate in contrast to the lines, wrinkles, and inscriptions of the face. "The excess of display turns everything into a commodity;

possessing 'no secret'," Byung-Chul Han (2015, 11) wrote in reference to Benjamin and Baudrillard, "instead they fade into the more visible than visible, obscenity." *Superficies* are obscene, because they are slick. They do not open a scene that is an open space for encounter. Thus seen, avatars and robots, despite their mimetic perfection, are genuinely *faceless*.

8. THE ETHICS OF RELATIONS

Could we then mistake machines for humans? Is possible to have a utopia or dystopia of a world inhabited by artificial machines and programs that we meet as equals, as fellow humans, because we are unable to distinguish between the two? Our explorations have led us to admit that such deception is possible, but only as long as we focus on perception and thus on similarities and dissimilarities. In that case, however, we would remain within the framework of traditional metaphysics or ontology, as does so-called posthumanist speculative realism. If instead we focus on the fact that robots and avatars, no matter how ideal their appearance, in the moment of their greatest similarity with humans, hide exactly that which would first make human relations with them possible; we come to another conclusion. Put another way, the same metaphysics and ontology is found in the ideal of a replication of the human form and these efforts toward best possible convergence can only fail because they lack that which Levinas apostrophized as the transcendence of the face, the site of ethical relations between humans. Concurrently, they lack the resistance and the inhibitions that go along with the singularity of the person, their unviability and vulnerability, categories that have always been cited in ethics. Therefore, our relations as *relationships*, as *relating*, have a fundamentally ethical basis that begins with perception and continues in the image. When in contrast we deal with calculations rather than thoughts, with operations rather than practices, and with repetitions rather than alterations, violent relations dominate. That is also the reason why we are able to show no mercy to robots and avatars and can enjoy films and computer games that very realistically blow up the avatars that we ourselves created to represent us.

For that reason too, robots and avatars cannot truly become objects of our desire, if under desire we understand the elementary craving or longing and care *for* otherness. Desire then also means, to vouch for the other, for he or her is the condition of possibility that "I" am. There is no similar responsibility for technical things or artificial life because we are not existentially conditioned by them. We must however delineate this more closely. Pygmalion's impossible love for his creation belongs in the field of cautionary fables and mirrors the phantasmatic nature of its narcissism. The story advises against its monstrosity. The same applies to the avatar as an object of identification and as a proxy figure. It can act as an egologic *projection* of our moves in play or as a *fetish*, but it disappears the moment the game is over. In short, technical artifacts may take the place of "actors" and we may accept this by giving them temporal emotions and looking at them like children. And they may also be superior to us in many realms, in particular those in which we do not want to compete with them in any meaningful way. But these remain attitudes that recognizes them as partners in participation, bound in the idea of a network of interests. The privileged rationale of this network is interdependence that always distinguishes *intentio* in order to unravel the knots of varying options for relationships. More interesting are the spaces in between the stiches. This emptiness, this place where no relations have been mapped out, is the site of our interpersonal relationships. This means there are at least two kinds of relationships, two different modes of relations, one for things and artifacts derived from human interactions, the other for human face-to-faces correlations, which are based on a primordial "*passibilité,*" which first and foremost enable an active encounter. This always already contains an ethical connotation and occurs from a space of reception. Both are not symmetrical. That is why Gilles Deleuze and Felix Guattari (2003, 168) called the face a "black hole" that sucks the gaze and desire into its center of gravity. This fall induces a *conversion*. It assumes the "gift" of the face. No avatar, no technical object, and no apparatus, despite its role as actor, can be a gift in this sense and in the sense of a singularity—which in turn is only a gift where there is inexchangeability. Artifacts, whether they speak or act, present only white walls rather than black holes. And it is because of this flipping of relations, this core asymmetry or incompatibility that we, in the sense of a primary ethicalness, respond differently to the "human countenance" and to the "facial image" of robots or avatars.

NOTES

This chapter was translated by Laura Radosh.

1. A pointed description of how we can be fooled can be found in E. T. A. Hoffmann (2008).

2. Translator's note, In German, "to reference" or "to refer to" (*sich beziehen*) is a transitive verb with the same root as "relationship" (*Beziehung*). Thus here and in the following, our manner of relating is, in fact, our reference system.

3. On the concept of the aura see also Mersch (2002).

4. On interpreting the philosophy of Levinas with a focus on responsiveness see Waldenfels (2007).

5. See in this volume chapter 2, Mira Fliescher "Digital Exchanges. Ghosts and the Gifts."

6. German, *Wendung des Bezugs*, see Mersch (2010, 287ff.).

7. Also related is the word *anthr-ōpon*, human or human countenance.

8. In late Hellenistic theater *prosopeion* stood for the mask and *autoprosopos* for the face that showed itself.

9. In digital facial rendering, data about and "landmarks" of the face are first captured and pictured before they are rasterized and modeled using varying geometric functions to simulate a recognizable image. Typically, the basis of facial rendering is an image of the face. In constructing avatars, this procedure is reversed, in that the "face" that was created from measuring and quantifying an image is first generated and modeled using mathematical values. This dual pictoriality turns the "facial image" (*superficies*) of the avatar into an anesthetic apparition. It is analytical—not phenomenal.

10. The term "operation" rather than "action" or "practice" is relevant here. See Mersch (2016, 31–52).

11. The terms "nature" and "naturalness" are used here only to refer to that which circumvents the technical. Since Donna Haraway's introduction of "natureculture," these two poles have been all too quickly amalgamated, brushing aside the possibility of an "Other of ourselves." But not only does the bulk of the human organism remain beyond technical control despite all medial advances and talk of cyborgization, technology can also only transform that which the "earth" offers by way of materials. Our remains and legacies that which is left over, return to earth to be repressed and rot. Without a doubt, the moon and stars have been shaped by culture and symbolically determined, yet something remains in their orbit that evades our grasp. The universe is so far from reach that it is indifferent to all fantasies about "natureculture."

12. In truth, this unpredictability is nothing more than the product of a randomization based on calculations.

REFERENCES

Balázs, Béla. 2010. *Early Film Theory*, trans. Rodney Livingstone. New York, Berghahn Books.

Bataille, George. 2002. "The Mask," trans. Jeremy Biles, *LVNG 10, Translations*, 63–67

Beil, Benjamin. 2012. *Avatarbilder. Zur Bildlichkeit des zeitgenössischen Computerspiels*. Bielefeld, transcript.

Belting, Hans. 2013. *Faces. Eine Geschichte des Gesichtes*. Munich, C.H. Beck.

Benjamin, Walter. 1997. *A Lyric Poet in the Era of High Capitalism*, trans. Harry Zohn. London, Verso.

Benjamin, Walter. 2010. "The Work of Art in the Age of Its Technological Reproducibility (first version)," trans. Michel W. Jennings, *Grey Room 39* (Spring 2010), 10–37.

Caillois, Roger. 2001. *Man, Play and Games,* trans. Meyer Barash. Urbana, University of Illinois Press.

Calvillo-Gámes and Cairns, Paul. 2008. "Pulling the Strings. A Theory of Puppetry for the Gaming Experience," in *Conference Proceedings of the Philosophy of Computer Games 2008,* eds. Stephan Günzel, Michael Liebe, Dieter Mersch. Potsdam, Potsdam University Press.

Deleuze, Gilles and Guattari, Felix. 2003. *A Thousand Plateaus*, trans. Brian Massumi. London, Continuum.

Han, Byung-Chul. 2015. *The Transparency Society*. Stanford, Stanford University Press.

Harman, Graham. 2007. "On Vicarious Causation," in *Collapse II,* eds. Robin Mackay. Falmouth, Urbanomic, 171–205.

Hoffmann, Ernst Theodor Amadeus. 2008. "The Sandman," in Hoffmann, *Two Mysterious Tales*, trans. John Oxenford. New York, Mondial, 3–42.

Hayles, N. Katherine. 1999. *How We Became Posthuman, Virtual Bodies in Cybernetics, Literature, and Informatics*. Chicago, University of Chicago Press.

Levinas, Emmanuel. 1969. *Totality and Infinity*, trans. Alfonso Lingis. Dordrecht, Kluwer Academic Publishers.

Levinas, Emmanuel. 1986. "The Trace of the Other," trans. Alfonso Lingis, in *Deconstruction in Context*, eds. Mark Taylor. Chicago, University of Chicago Press, 345–359.

Levinas, Emmanuel. 1998. "The Ruin of Representation," in *Discovering Existence with Husserl*, eds. Emmanuel Levinas, Richard A. Cohen and Michael B. Smith. Evanston, Northwestern University Press, 111–121.

Matt, Peter von. 1983. *. . . fertig ist das Angesicht. Zur Literaturgeschichte des menschlichen Gesichts*. Munich, dtv.

Mersch, Dieter. 2002. *Ereignis und Aura Untersuchungen zu einer Ästhetik des Performativen*. Frankfurt/M, Suhrkamp.

Mersch, Dieter. 2010. *Posthermeneutik*. Berlin, Akademie Verlag.

Mersch, Dieter. 2013. "Turing-Test oder das "Fleisch" der Maschine," in *Körper des Denkens*, eds. Lorenz Engell, Frank Hartmann and Christiane Voss. Munich, Fink, 9–28.

Mersch, Dieter. 2016. "Kritik der Operativität," *Jahrbuch für Medienphilosophie Band 2*. Berlin, De Gruyter, 31–52.

Mori, Masahiro. 2012. "The Uncanny Valley," trans. Karl F. MacDorman and Norri Kageki. Posted online in *IEEE Spectrum* on 12 June 2012, http:// spectrum.ieee.org/automaton/robotics/humanoids/the-uncanny-valley (Last accessed 24 February 2016).

Morton, Timothy. 2008. *Ecology without Nature. Rethinking Environmental Aesthetics*. Cambridge, Harvard University Press.

Morton, Timothy. 2010. *The Ecological Thought*. Cambridge, Harvard University Press.

Turing, Alan M. 1950. "Computing Machinery and Intelligence," *Mind* 49, 433–460.

Waldenfels, Bernhard. 2007. *Antwortregister*. Frankfurt/M, Suhrkamp.

Weihe, Richard. 2004. *Die Paradoxie der Maske. Geschichte einer Form*. Munich, Fink.

Wesserly, Christian. 1997. *Von Star Wars, Ultima und Doom*. Frankfurt/M, Peter Lang.

Chapter 2

Digital Exchanges

Ghosts and Gifts

Mira Fliescher

1. DIGITAL EXCHANGES

The economy of communicating with others on social media is a signature of the digital age. On social media, one acquires a face by *giving, disseminating,* and *sharing* information, comments, and "likes." People not only share information about themselves, but also about others. Social networks function by soliciting people to make public both their own lives and those of others. This phenomenon has occasioned serious reflection on how to keep the private sphere secure. Even if one could write off the whole exchange of "shares" as a cheap way to collect data (i.e. for market research), the question as to what sort of sociality is fostered by social networks remains relevant, and with it, so does the related question as to what sort of ethics they foster. Is it possible to conceive of the exchange of "shares" as a form of communication that goes beyond mere data traffic, opening space for a relation of responsivity to the Other?

This communication primarily takes the form of posts, comments, opinions, and messages. In turn, social media regulate communication as an exchange of *data.* With names like *Facebook* (of all things), the programs and algorithms that run these applications regulate and restrict the possibilities of exchange, the forms of subjectivity that can be constituted on them, and the actions that can be performed through them. Recently, the politician, Internet activist and blogger Katharina Nocun put forth the demand that social media networks be democratized by making it possible for users to choose their preferred network while at the same time being able to interact with others on all other social media platforms, in a way similar to e-mail (Weber 2015). This would supposedly make it easier for users to skirt the limitations imposed by the various social media services, and it would help users have more say in how services treat the data they collect. In the end, however, users would really be able to choose only the user interface and the company that exploits their data. Because making the information of users on one network accessible to users of another would require the various services to be compatible with one another. That is, they would have to become equivalent in some ways. But the forms of communication that they enable would still have to be determined in advance before the question of overarching compatibility could be tackled. For instance, the services would still have to decide in advance whether or not there should be a "dislike" button or only "likes," or, in the newest development of things, which emojis can be used to respond to a comment. This sort of democratization has already integrated the limits of Internet communication into its very structure; its way of providing equal opportunities is itself determined by the indifference of "the program." Thus, while such demands might be necessary for the politics of the everyday, they nevertheless only grapple with the political and legal regulation of user choices, which is ultimately a secondary matter. They therefore fail to address the ethical and social dimensions of the phenomenon, which, in the philosophy of Emmanuel Levinas, have their beginnings in the response to the Other.

Three observations can help us grasp why the ethics of alterity is relevant for our understanding of communication by way of social media:

(1) The sociality of social media begins with a representation of *oneself*, a requirement that is pre-scribed by the programs of social networks. Only with an account or identity (real or fake) can one find one's friends. This very *structure* determines the sociality of social media interfaces in advance: it begins with "I," because without "I" there can be no encounters with other "I's." Byung-Chul Han has written about the trend of people posting pictures of themselves on *Instagram*, claiming that it evidences the liquidation of "seeing." The aura of the face that Walter Benjamin spoke of in his analysis of portrait photography is completely absent from these photos, such that they, according to Han, only have value insofar as they put their subject on display. He thus writes: "People comport themselves like commodities" (Han 2016).

Rather, one might say that they comport themselves in conformity with the computer program, because the apparatus and the program are what structure this narcissistic egology that conceals its own commodity form from itself. The program says: "Connect with friends and the world around you on Facebook." It asks: "Forgot your password? What's on your mind?" It asks you to: "Sign up. Log in. Complete your profile." It suggests: "People you might know." And then the cycle begins: "Like. Comment. Share." Thus, it is the *programs* themselves that prompt one to practice egology in the first place. First, you have to give *yourself* a face that conforms to predetermined parameters (even if it's just the default silhouette), *then* you can post.

(2) Thus, the sociality of social media consists in communication between one self-identified "I" and other "I's." But in criticizing the spectacle of posing on *Instagram* as a form of self-alienation, Han misses something. No doubt, taking a picture of oneself jumping—a pose whose very "activity" gives it a veneer of authenticity—is the product of a failed attempt to stick out in the pre-programmed world of mass self-representation. But attempting to outdo and undermine the simulated character of social media self-representation is symptomatic of the logic of social media itself. What is more, Han himself submits to the alienated egology of social media by implicitly demanding that these media should foster a form of sociality where the self-representations of singular "I's" would give him access to an Other that conforms to his idea of what or who the Other is. One way or the other, the "face of the Other" cannot reveal itself in pictures of people jumping, because the images used to identify me on social media are, in the end, only an instrumentalized version of the face, an "I" that I have to invent in order to gain publicity. This is why Han neglects to write about responsibility, remaining satisfied with criticizing the ways others represent themselves.

(3) Digital communication fosters a forced form of criticism that takes the guise of "shitstorms" and other disturbingly harsh remarks. Such expressions of disrespect proliferate without restraint or empathy, which is all the more astonishing when compared with face-to-face encounters and the culture of writing. It is as if the structure of the "post-share-repeat" cycle has taken on a life of its own. Lacking responsibility, recognition of the vulnerable Other being addressed, and any consideration of the effects that remarks might have outside the social network, it seems that users have made the act of posting and its

over-the-top expressions into ends in themselves. Even when something has lost its relevance, even when someone who started a shitstorm has apologized, even if the whole conflict was just a misunderstanding, even when two squabblers have set aside their dispute; posts continue to be quoted, shared, commented on, and liked, as if they had neither an addressee nor any effects on the lives of others. The mere formalism of posting and the compulsion to outdo others seem to absolve users of their sense of responsibility. Exchange becomes an end in itself in the potlatch of posts. The formalism of the drive to outdo others in the extremity of one's posts calls for a renewed inquiry into the ethics of the gift. Can social media mediate or give the face of the Other? Do they have the capacity to allow the vulnerability of the Other to break through the formalism of posting for posting's sake?

In 1925, Marcel Mauss (1966) claimed that gift exchange lay at the foundation of human sociality. Like Bataille's (1988) later analysis of excess, Mauss's concept of potlatch was not aimed at providing a financial or economic explanation of sociality, simply because in the gift-giving festival of the potlatch, the requirement that every gift be reciprocated is taken to the point where utility and economic efficiency give way to the general ruin and destruction of goods. This, however, can have a symbolic value; for instance, by securing the power of a leader or humiliating others (Mauss 1966, 93). That which Mauss called the gift is not a mere commodity, but a symbolic good. Pierre Bourdieu (1998) later criticized this conception as an obfuscation of commodity exchange that merely transfigures it into something symbolic, accusing Mauss of deploying the guise of the symbolic in order to turn sociality into a business venture. Nevertheless, this critique doesn't help us much in our inquiry into the ethics of the gift, as it does little more than moralize the structure of exchange. For this reason, Jacques Derrida (1992, 7) distinguished between gift and exchange:

> If there is gift, the *given* of the gift (*that which* one gives, *that which* is given, the gift as given thing or as act of donation) must not come back to the giving . . . It must not circulate, . . . it must not in any case be exhausted, as a gift, by the process of exchange, by the movement of circulation of the circle in the form of return to the point of departure.

In order that there be a real gift, "the gift must remain *aneconomic*. . . . It must *keep* a relation of foreignness to the circle, a relation without relation of familiar foreignness" (Derrida 1992, 7). The gift is an alterity that ruptures the economy of circulation (which, for Derrida, encompasses both the exchange of commodities as well as that of signs). The gift is paradoxical, because in a certain sense it cannot be given. "It is perhaps in this sense that the gift is the impossible," Derrida (1992, 7) writes. "Derrida (1992, 7) writes: "Not impossible but . . . the very figure of the impossible." The gift thus poses a limit to circulation. This limit is important for an ethics of social media networks, because the gift introduces an ethical break with symbolic circulation. Moreover, it does so within a certain *culture of writing*, which we originally placed in opposition to the potlatch of posting. Does this mean that social networks are analogous to writing and that the anonymized inconsiderateness shown toward the addressee is a horrific transformation of the gift of writing? Then writing would be no different than a program and its gift an ethically inefficacious poison.

So why does Derrida think that the circulation of exchange has to be broken? In order to address this question, I would like to give a short overview of the way in which Derrida developed an ethics out of the concept of writing. I will focus on two points: (1) The limitless, almost automatic iterability of writing constitutes an inhuman dimension of communication that Derrida views as horrific the moment it takes on a life of its own and enters into a self-regulating cycle. (2) It is precisely at this juncture that deconstruction recognized the necessity of an ethics of writing. And yet, this ethics cannot constitute itself in the structure of writing alone. This ethics requires an impossible outside of writing. This becomes all the more true when writing is conceived of as a mechanism or code.

2. GIFTS OF LIMITED INKS

In "Signature Event Context" Derrida uses his theory of writing to explore the problem of communication. He criticizes the Shannon-Weaver model of communication because it implies that a fixed message is transmitted from one consciousness to another. This model of communication fixes meaning by centering it, in egological fashion, in the self-present consciousness of a subject. Derrida deconstructs this model by drawing attention to the structurally necessary absence of the signer and the addressee in writing. He goes on to note the lack of disposability over the medium of writing in every utterance. Writing cannot be grasped before, during, or after the act of enunciation (Derrida 1988a). Derrida explains the writer's constitutive lack of disposability over the medium of writing by analyzing the phenomenon of the signature. The signature, as sign of the "I," is supposed to function as a self-authorization: it has economic and legal consequences (i.e. contracts), it seals the intent of the signee as well as the meaning of her utterances, and it serves as an expression of the irreproducible of individuality. However, in order to fulfill all these functions, the signature has to be valid, which demands that it be recognizable, that its original be documented somewhere, and that it transmit a sameness. Thus, in order to represent consciousness, the signature *as sign* must be repeatable (this is why many people practice their signatures). It aims at producing identity, but can only do so by being different; it is always countersigned by writing as such and by institutions (Wirth 2007, 66). Thus, the act of giving one's signature is *preceded*, *permeated*, and *succeeded* by the constitutive and irretrievable alterity of writing's peculiar logic, which decenters intentionality as the metaphysical center of meaning. This is due to the fact that *différance*—the force of writing—constitutes sense while at the same time disseminating, differentiating, and deferring it. The signature thus tags a written *trace* that proceeds in an infinite displacement of differentiality without any embodiment or presence. Read as the definitive feature of writing's mediality, *différance* marks an *alterity* that perpetually withdraws itself while dispensing with any sort of *aesthesis* or materiality. Writing *traces* without showing itself—in a non-material, *anaesthetic*, self-inscribing movement.

"Signature Event Context" is significant for our inquiry into a possible ethics of digital communication because it undermines the notion that communication consists in the transmission and reception of messages through signs, that is, that it consists in the transmission of information. The text disturbs the logo-phono-centric localization of sense by showing that communicative intentionality is always contaminated by the mediality of writing as the constitutive Other of consciousness. So, the medium of writing—which, it should be noted, is not a program—necessitates a representational egology that is at once undermined by the structural absence of addressor and addressee in writing. Interestingly enough, this structural absence can also be found in the egology of social media. One could be tempted to understand social media as a means of turning the concept of writing into a program by turning *certain aspects* of writing into the very *principle* of Internet communication. One might also say that the *copy and paste* of sharing, posting and reposting is permeated by iterability. If this were true, then the idea of a stable message circulating in communication would be disrupted by the differential iterability of the sign:

> As far as the internal semiotic context is concerned, the force of the rupture is no less important: by virtue of its essential iterability, a written syntagma can always be detached from the chain in which it is inserted or given without causing it to lose all possibility of functioning, if not all possibility of "communicating," precisely. One can perhaps come to recognize other possibilities in it by inscribing it or *grafting* it onto other chains. No context can entirely enclose it. Nor any code, the code here being both the possibility and impossibility of writing, of its essential iterability (repetition/alterity). (Derrida 1988a, 9)

"Every sign, linguistic or nonlinguistic, spoken or written . . . can be *cited*, put between quotation marks; in doing so it can break with every given context" (Derrida 1988a, 12). Thus, it is the normal functioning of the sign itself that makes it possible for deconstruction to transgress—through the "logic that ties repetition to alterity" (Derrida 1988a, 7)—the limitations imposed on meaning by the code, which constitutes at once "the possibility and impossibility" of writing. When the logic of exchange pre-scribed by a program or a code takes on a life of its own, like in social media shitstorms, it is no longer clear how writing differentiates itself from the calculus of the code or how it allows an otherness to appear through the rolling tide of *copy and paste*. Whatever finds its way into the writing machine of social media *perpetuates* itself in a seemingly automated, purely iterative fashion. While this movement has similarities with writing's dethronement of intentional consciousness, in repeating the ever same, it lacks the dimension of alterity.

Friedrich A. Kittler took up Derrida's theory of writing in order to "banish" "the human from the human sciences" (Kittler 1980, 7–14), a despiritualization that reaches its zenith in the computer program. Kittler's claim that "both people and computers are . . . run by programs" (Kittler 1999, 17) thus contains a "masked *a priori* of writing" (Mersch 2005, 198). This *a priori* implies that the programmatic structure holds for writing as well, which becomes all the more pressing when one considers the fact that the early Derrida thought that there is no outside of writing, because writing always already structures every consciousness and every intentionality. But it is precisely at this juncture that the grammatological critique of intentional egology runs the danger of universalizing writing in a way that would erase every form of alterity. No praxis, no event, no Other would be able to come in from the outside. Everything would become an inevitable effect of a writing that would write itself along as the procession of that soulless monstrosity that is the writing machine, conditioning everything in advance. With more or less explicit reference to the works of Kittler, German Media Theory has advocated conceiving of media as an *a priori* in an analogous way, even when theoreticians take note of the parallels between cybernetics and writing with the aim of turning this condition of the medium into a *pharmakon* that would enable a new form of media praxis (Hörl 2005). In the end, however, this antidote reduces writing to code.[1] Writing is interpreted from the perspective of the computer code and its programs rather than the other way around, such that the radical alterity of writing as well as the radical alterity of the Other are *calculated away*. No doubt, "Signature Event Context," with Derridas attempt to demonstrate the evasiveness of writing by "giving" three versions of his "signature" at the end of the essay, may have paved the way for this technologically reduced concept of writing.

However, the effects of this text forced deconstruction to reflect on a problem of ethics. The debate that ensued between John R. Searle and Derrida over the latter's interpretation of John L. Austin's linguistic theory and the seriousness of deconstruction showed that polemic, critique, and misattributions could still have an impact on Derrida himself, who, despite all differential disseminations of the signature, remains the responsible author-signee of "Signature Event Context." In this "tracing" of the signature, a problem of ethics arose, because Derrida found himself in the uncomfortable position of having to defend his essay against the criticism that his reading of Austin was a misreading. Thus, Derrida seemingly became a victim of the same iterability of the sign that he had claimed was a definitive aspect of writing. In retrospect, it seems that the possible horror of an infinite iterability of the sign is the terrible underside of Derrida's decentering of the "I" and its intentions by means of the repeatability that undermines them. Unescapable and without end, this iterability bears the possibility of not only opening up new meanings but also letting injuries circulate. Circulation can thus become a purely formal process, similar to the processing of code or the formalism of transmission, giving permanence to injuries by perpetuating them. The debate showed that deconstruction cannot possibly take the constitutive absence of addressor and addressee in writing to mean that the Other who has written something remains wholly autonomous from the "tracing" bound to the signature. Rather, the signature is binding in two ways. The act of signing binds the Other to writing and to everything that has ever been written. *And*, writing and the trace occasion ascriptions that bring forth the signatory as a subject responsible for all of the effects of the readings of that which she signed. The "I" who writes is thus exposed to all of these countersignatures.

In his response to Searle, "Limited Inc a b c," Derrida indulges in the play of quoting his own work. In order to avoid making an egological defense of his original intention, the response makes a plea for a careful reading of texts by re-reading "Signature Event Context." Framed by the repeated demand "Let's be serious" and the question "Who, me?" (Derrida 1988b, 31, 34, 39 and 65), Searle's authorship is attributed to the various members (from Austin to those thanked at the beginning of Searle's essay to Derrida himself) of a limited liability corporation (in French, S.à.r.l.) in order to show that Searle has a share in the meaning of "Signature Event Context" and that Derrida has a share in Searle's critique. In this way, all of the shareholders bear responsibility for the *circulation* of the effects of reading, while at the same time, the subject made responsible as signatory takes refuge in the withdrawal of writing and the plural disseminations of the signature. Whereas Derrida had at first attempted to use the concept of iterability in order to deconstruct—or even *to avoid*—the problems of egology, he now discovers that an ethics of writing was lacking, a consequence of the absence of the subject in writing that could only be resolved through this very absence itself. Because only then could an ethics of writing allow an Other to be written as alterity, rather than merely citing others as subjects.

The response is thus to let the impossibility of the gift of the Other *play itself out* through the alterity of writing.[2] The pluralization of the signature's referent through writing, and the signature as marker of the impossibility of the relation to the signatory as Other, shield against false objectifications that can never do justice to the Other qua Other. Nevertheless, beyond all identifications and objectifications, there has to be a gift of the Other that gives the Other *in absentia* through writing. This gift of the Other breaks through the circulation of signs, in order to push the traces that are just writing along on their way in a new direction. The insight that something has to disrupt the representational, communicative economy of the mere exchange of signs is derived from the deconstruction of egology through writing: not because writing lacks an intentional ego, but because it threatens to put the automated violence of iteration (as opposed to iterability) in the "I's" place. Thus, there must be a relatedness with the Other in writing that is not a purely formal relation. Its non-formality is indicated in the impossibility of the reciprocity of the gift. This does not mean that there is no exchange. It just means that something must go beyond the formality of the swap: the gift "must *keep* a relation of foreignness to the circle" (Derrida 1992, 7).

Might this not also be possible in social media networks? The Other's "post" comes to me as a trace of this Other and I am encouraged to respond with "like" or a "comment." Does the Other not call me anyway through the codes and programs? Yes and no, because the program constantly puts the same formal structure in place, both for me and for the other: here, "I" write like the other person and the other person writes like "I" do. The program has already formalized both in the same way. The program thus not only pre-scribes a sort of stimulus-response schema of "pushing buttons," but it also pre-scribes an "I"-form of the Other that establishes the alterity of the Other as an identity defined in-line with its parameters. Even if the pre-scribed markers of this identity remain (for the most part) "empty," they are nevertheless *programmatic* variables; not even an absence, but only a *lack of information* without any mystery or the gift of the Other that might rupture the structure of the code. The recursivity of the network does not generate a surplus out of its lack, and a computer program cannot do anything but propagate exchange without remainder. Otherwise, it would break down. Its algorithms preclude the paradoxical. And yet, the rupture of the Other seems to instantiate itself precisely in the form of the paradoxical. However, writing has—at least for Derrida—the potential to harbor paradox. In writing, the aneconomic nature of the gift ruptures the circular economy of exchange. In its foreignness to exchange, it produces a surplus. And yet, at the same time, because writing leaves no room for alternatives and has no outside, the gift must remain just as imperceptible and unnoticed as writing itself. Thus, the gift is an inner externality, an impossibility that oscillates between inside and outside, impossibility and possibility in order to make an outside thinkable. Through this constitutive aporia of alterity, deconstruction seeks to challenge its own limitations (García Düttmann 2008, 13ff.). The gift establishes a break that challenges the circulation of signs in writing in order to change the trajectory of the trace. The rupture of the gift thus sets a limit to the automatism of writing, a limit that has to remain foreign to it. This limit is, at least initially, the act of giving, a praxis of time, a deferral that allows an opening that counteracts the circulating stagnation of the ever same.

3. NON-SUBSTITUTABILITY (LIMIT II)

However, if this paradoxically pluralizing "limited ink" is to fulfill its promise of an ethics of writing that is responsive to the Other, then it has to meet another challenge. This responsibility necessitates that deconstruction submit writing to a further limit: The Other has to be introduced not only as pluralized absence, but also as a singular Other characterized by an unsubstitutable vulnerability. This limit is given by the non-substitutability of the death of the Other. This limit is an impossible limit because death is the non-representable site of the absolute singularity of the Other. Thus, Derrida (1995, 42) quotes Heidegger (1977, §47, 240 [284]): "No one can take the Other's dying away from him (Keiner kann dem Anderen sein Sterben abnehmen)." To that, Derrida lays particular weight on Levinas's insight that death signifies the absolute abruption of responsivity. Alongside his discussion of the Levinasian figures of responsivity, substitution, duty, and the absolute alterity of the Other, Derrida develops an ethics that is founded in death as the final, irrevocable rupture. According to this ethics, the impossibility of the gift is grounded in the absolute impossibility of substituting for the other. It thus precedes every formal morality. This ethics is always already guilty, yet it does not owe anyone anything, because there is no possibility of giving anything back. There is no possibility of any kind of re-presentation. The (possibility of the) Other's death preserves the Other's vulnerable singularity as mystery and irrevocable withdrawal that can be neither shared nor communicated. This radical alterity inevitably makes the relation to the Other an asymmetrical relation. For Derrida, this asymmetry is not derived from God, but from human finality, and it must inscribe itself in every response. In grounding his ethics in a truly aneconomic gift, Derrida brushes aside every egological ethics that disguises itself as altruism. But in "The Gift of

Death," he places his focus on an ethical duty that, with clear reference to Levinas, regards the radical, singular alterity of the Other. However, by conceiving of the death of the Other as non-substitutability, he implicitly alters Levinas's notion of substitution.

Levinas uses the concept of substitution to describe the recurrence that follows *after* contact with the Other. Substitution ultimately results in the constitution of an *I* that is at once responsible and non-egological. The substitution of "one-for-the-other" (Levinas 1998, 158) denotes a standing *in the place of* the Other, which *follows out of* the an-archic, passive obsession with the other after the epiphany of the face of the Other. This means that substitution "has already happened" (Bernasconi 2002, 250). In the same way that the Other always comes first for Levinas, the Other is already there before an I as "hostage of the Other" becomes irreplaceable by way of standing in place for the Other, that is, by way of responsivity: "The I, the unique one, substitutes itself for others" (Levinas 1998, 117). Conceived of as "the possibility of putting oneself in the place of the other," substitution implies an impossibility of exchange (Levinas 1998, 117) that undermines every purely formal relation to the Other. Under the spell of the Other, this asymmetrical relatedness gives rise to a surplus of responsibility. This surplus has nothing to do with any kind of calculation, but rather stems from the impossibility of trading places with the Other's unique vulnerability.

While many instances of Internet hate speech have had very real effects on the lives of those attacked—with some cases even ending in the death of others—it seems that many of the judgments made on social media neither consider nor instantiate this impossible exchange. The fact that comments and posts can quickly cause one to lose one's social face (or one's job) is not merely a result of the omnipresence of the Internet. It is more indicative of a total lack of consideration, carried out without the least bit of empathy, forgiveness, or ability to look past the flaws of others: from the comments about the tasteless fashion choices of a celebrity (who at least profits from the publicity, even if it's bad) or a neighbor, all the way to cyberstalking and bullying. Self-representation and the affairs of others have value only insofar as they can be exhibited in a way that gives the "I" who posts a chance to milk them for (social) capital by indulging in scandal. Thus, this egology returns to the "I" at the expense of others without the slightest bit of alterity. Even worse, friendly gestures toward others (happy birthday wishes, congratulations, etc.) have to be tallied up on the self's account.

They are always first and foremost a part of one's own profile, which not only counts how many friends one has, but also keeps track of the number of posts one makes. Thus, every single post starts and ends by contributing to the program's I-producing apparatus.

In his extensive textual exegeses, by contrast, Derrida seems to deploy writing as the residuum of a non-egological position that is at once active and passive; as something like a writing machine that undermines all egologies of writing through the iterability of the sign in order to generate a surplus of alterity that might make room for the Other. Thus, Levinas's concept of substitution—which denotes something that has always already happened, that is passive, and that is ultimately impossible—motivates Derrida's later writings. In inscribing death, the rupture of communication, the limit of writing into writing itself, Derrida's writing practice wants to be affected by the vulnerable unsubstitutability of the Other without returning the gift and without any attempt of representation. Substitution thus becomes the task of a *gift of writing* and a *gift in writing*. Where Levinas posited an absolute and indivisible oneness, this practice of writing divides the singularity of the Other. Further, the radical passivity of substitution is turned into the withdrawal of the gift: it is at once the gift of writing and a giving of writing in which the Other *and* something other always write along. In developing his theses in lengthy responses to the writings of others, at stake in Derrida's writing is not only a deconstruction of established meanings but also the question of a responsive mode of writing. Its impossible gift of a "response without response" (Derrida 1995, 75) is given as something that is touched by the Other and that is supposed to give refuge to the Other in absence.

Writing is thus called upon to *substitute* for the impossibility of giving an unsubstitutable substitutability. It is able to do so precisely because it is an alterity which withdraws from the ego in its auto-logical generation of non-identical iterations. Writing bears the asymmetry of the relation to the Other because it is unable to remunerate the Other. And because the Other has no positive presence in it, it is possible that the Other "alter-nates" with a writing that marks the absence, the lack, and the unsubstitutability of the Other. Through the event of a performative responsivity, this possible impossibility bears writing's promise to transcend the code and the code's guarantee of self-identical repetitions. Thus, writing must be more than code; there must be an *alter* in the *iter* (Derrida 1988a, 9). And yet, this event of *alternation* can only take place within the process of writing. Thus, it has to be constantly differentiated through new figures (*différance*, gift, death). The promise of a rupture of the circle has to be renewed by writing over and over while at the same time postponing this rupture. The question is, therefore: Is there a code at work here that regulates the identical return of the same figure in different guises? Can the face of the Other appear, or is it all just a continuous recourse of writing to writing, which would really be nothing more than a code? These questions are even more pressing for digital communication, because it sometimes confronts us with pre-programmed "others" "who" disguise the program's set-up in personalized masks. Using human, animal, or purely fictional sur-faces, they imitate faciality without being able to have a face. This attempt of the code to humanize itself robs users of their capacity to recognize whether they are addressing a vulnerable Other or an inhuman interface. And at the same time, it makes it impossible for users to recognize if they themselves are acting like a dehumanized function of the program. Thus, one is confronted with irritating demands like having to solve a Captcha puzzle in order to prove that one is not a robot. One is asked to prove one's humanity by submitting to the parameters of a program. If deconstruction's ethics of non-substitutability is to make an impact here, it must clarify once and for all whether its concept of writing and non-substitutability makes it possible for the face of the Other to appear or whether it just perpetuates a *programmed script*.

4. WITHOUT EPIPHANIES

Derrida often reads absent presences of the Other in others' writings. For instance, in *The Gift of Death* he reads Heidegger's "not seen" present in Patočka's work (Derrida 1995, 39), and in "Typewriter Ribbon: Limited Ink (2)" he is able to "hear . . . already" the missing objections of Paul de Man. In "Typewriter Ribbon," Derrida interprets de Man's concept of the textual event as a threat to the subject (Derrida 2002, 154ff.), insofar as "the text as machine" (de Man 1979, 297–298) harbors the possibility of a loss of reference and meaning. De Man interprets this as a sort of "machinic" materiality that would proceed without desire, randomly, automatically—but not "programmed" (Derrida 2002, 158). This "machine" produces the textual event as a moment that confronts us with the inappropriability of the text. In agreeing with the relevance of such a textual event, Derrida insists that this event contains a traumatic encounter with the Other. It is "some outside, . . . the becoming possible of the impossible *as* im-possible. Inappropriability of the other" (Derrida 2002, 159–160). Thus, for Derrida, the Other *haunts* the machine. The textual event as trauma comes *from the Other*; it bears witness to the absence of the Other and of a missing Other. This is why Derrida rejects de Man's (ironic) statement that: "Whatever happens in Derrida, it happens between him and his own text . . . He doesn't need anybody else." (de Man quoted in Derrida 2002, 160) Derrida believes that he can already hear de Man's responses, hoping that "sooner or later [de Man's] text will answer for him" (Derrida 2002, 160). He writes: "That is what we all call a machine. But a spectral machine" (Derrida 2002, 160). Thus, a spectral machine mediates the revenants of the Other and mediates in place of the Other. If that which "we all," according to Derrida, "call a machine" is the text as an (immaterial) materiality that constantly produces responses, then how to comprehend this impossible-possible machine that produces an event of responsivity bordering on the hallucinatory? Does not Derrida need the Other *for himself*? Is the face of the Other only there to justify the writing "I" by adorning its interpretations in altruistic gowns? While writing's deferral is supposed to make room for the Other, nothing can guarantee that the machine that threatens egology actually opens up a "spectral" response instead of just mechanically cawing along. Thus, no doubt for systematic reasons, it remains a mystery as to how the Other receives a face in writing.

Nevertheless, at least Derrida's writing constantly remains aware of the fact that it is doing more than just "pushing buttons." Derrida's texts always revolve around a certain impossibility of responsibility in writing. Responsibility has its origins in the fact that the signing/signed Other is bound, *in absentia*, to the given signature and to the ways it is co-signed by acts of interpretation, reading, and response. As such, the Other is *exposed* to writing's iterability, whose very structure is the cause of the proliferation of the "traces" of the signature. The irretrievable alterity of writing, absolutely necessary for the Other's call for responsivity, is made into the foundation of an ethics that exploits the structural analogy between the alterity of the Other and the alterity of writing as a sort of remainder of their difference. Derrida (1982, 6) writes that *différance*, the driving, performative force of writing, is "in every exposition . . . exposed to disappearing as disappearance."

Similarly, "the other can only appear by disappearing" (Derrida 2001, 48). They mutually preserve one another *by* withdrawing. It is only the withdraw *of* writing *along with* a withdraw *into* writing that gives the other a chance to be an Other *at all*. Writing does not archive some positive trace of the Other that could be hunted down by the detective work of a "police discourse" (Derrida 1987, 363), but only preserves undecidability. This mutual withdrawal also preserves the essential difference that opens up the writing machine for the entry of the Other.

Thus, there is a gift of non-substitutability that allows the absolute alterity of the Other to be substituted through the intrinsic absence of the Other in writing. The Other is just as singular and imperceptible as writing itself without being identical with it. In order to make the appearance of the face possible *in* and *against* writing, writing must be turned against itself. The structural parallels between *différance*, the gift, writing, and alterity/the Other in radical withdrawnness offers an impossibility of the gift of the Other *as absence*, working both with the automatism of writing and against it. As such, they make possible, *without intention*, this very impossibility. Derrida thus places his wagers on the chance of such an impossible substitution. However, as discussed above, this bet on the writing machine risks the undivided, absolute dimension of the alterity of the Other. On the one hand, the Other persists, as tautological self-sameness, in radical withdrawal, and on the other hand, the Other is localized in a series of pluralizations of others/othernesses (García Düttmann 2008, 13). The Other must necessarily remain a paradox for writing. Such a paradox can arise in relation to a code, but *as paradox* it cannot be codified or integrated. And yet, the paradox is not identical with the Other. Rather, the paradox *indicates* the fact that there is something that cannot be codified, quantified, or calculated. The paradox points toward the Other that necessarily appears without appearing, indicated by the substitution of writing. It thus remains undetermined as to *how* the "break-in" of the Other is able to reveal itself.

But Levinas has no need to pursue the question of "how," precisely because he conceives of the epiphany of the face or voice as a rupture that manifests itself in representation (or language) without motivation or structural possibility. The presence of the Other only realizes itself in absolute transcendence. It cannot be in any way perceptible or shown, because if it were experienced, it would already be *for* an intentional consciousness. The epiphany of the Other precedes every intentional

"I" and thus has the capacity to rupture all established meanings and phenomenal appearances. The being of the Other appears in its nakedness as a presence *without aesthesis*, *before* every intentionality and consciousness, *beyond* every mediation. Thus the emphasis on the face-to-face encounter. The Other befalls as an undivided singularity, even *without* having to withdraw into the deconstructive displacement of signifiers. The face enters the scene *without* even making any *contact* with these things. Because the face is a "naked existence," its "exposure" calls out for a response that (in substitution) shows itself responsible for the singularity and alterity of the Other. Because Levinas rejects any sort of mediation or *aesthesis* for the epiphany of the Other as a false contamination or division, he criticizes Derrida for seeking to "express this lack of presence positively" and thus for coming dangerously close to "returning to presence" (Levinas 1996, 60). Levinas suspects a mere formalism of writing and a return of intentional consciousness in Derrida's attempt to let writing become a substitute for the Other. In contrast, the early Derrida accuses Levinas of misunderstanding his own position: "What authorizes him to say 'infinitely other' if the infinitely other does not appear as such in the zone he calls the same, and which is the neutral level of transcendental description?" (Derrida 1978, 156). This friendly disagreement between Levinas and Derrida not only affects the status of substitution in their respective conceptions of ethics, but it also has bearing on Derrida's notion of an "*aesthesis* without *aesthesis*." In this context, non-substitutability and non-appearance are closely related to one another.

One reason for Levinas's ambiguous stance toward writing and *aesthesis* is that he suspects that the medium of writing and its roundabout descriptions become ends in themselves that erase the Other. And no doubt, the same concern spurred Derrida to make non-perceptibility an intrinsic feature of his notion of non-substitutability, because substitution is only possible between humans, or, for Levinas, between humans and God. The epiphany of the Other must remain an appearance without *aesthesis*, because only in this way can it serve to demarcate the holy difference between humans and everything else. With the question of the relation to the Other, Derrida thus distinguishes between two modes of imperceptibility. The first remains within the sphere of *aesthesis*. The second, however, completely transcends the *aesthetic* by persisting in a more or less godly "dissymmetry of the gaze," wholly withdrawn from any sort of direct encounter or representation; thus, it belongs to the sphere of belief (Derrida 1995, 29). This second mode corresponds, in a precarious way, with the withdrawal of *différance*. If one were not to conceive of writing as a sort of "deus ex machina," that is, as something distinguished from God, then writing would take the place of the absolute Other or enter into the domain of the human. That there is a gift and substitution only among equals (that is humans) is, in the end, an unspoken humanist condition shared by both Levinas and Derrida.[3] This condition establishes an element of symmetry within the asymmetrical relation to the Other, which is founded in a paradoxical "non-*aesthetics.*"

One can thus uncover the implicit humanism of both thinkers by analyzing their respective concepts of substitution: in Levinas, because of the way he founds the radical singularity of the Other exclusively in an im-possible substitution *between* humans; in Derrida, since he excludes any sort of materiality or *aesthesis* from his conception of non-substitutability, which, at least in writing, is made into an im-possible possibility or task. And this despite the fact that materiality and *aesthesis* are unavoidable aspects of Derrida's specific praxis of writing. Why do both rule out the possibility that the Other might be revealed in an appearance that would precede intentional consciousness? That such an appearance could never in itself give rise to an ethics or sociality is an entirely different matter. Because, of course, the Other must have been there *before* any epiphany of its traces can be mediated or show itself (on this point, see Dieter Mersch's remarks on the mask in this volume). Despite ironic comments like Derrida's claim that "materiality becomes a very useful generic name for all that resists appropriation" (Derrida 2002, 154), it becomes clear that neither Derrida nor Levinas are able to avoid an *aesthesis* for the Other. If the epiphany of the Other is an appearance *without* materiality and *without aesthesis*, then how is it possible to be aware of the Other at all? Why should one resort to audial or visual metaphors at all? Both thinkers remain undecided in their rejection of *aesthesis*. Thus, while in many places Levinas explicitly conceives of *aesthesis* as aiding a form of representation that addresses the intentional "I" through sensory images (Levinas 1987), some of Levinas's remarks run contrary to his general aniconism[4]:

> A graphologist, an expert in writing styles, or a psychoanalyst could interpret a trace's singular signifyingness, and seek in it the . . . intentions of him who delivered the message. But then what remains in the specific

sense a trace in the writing and style of the letter does not signal any of these intentions, . . . reveals and hides nothing. (Levinas 1986, 357)

For sure, there is a mediated trace of the Other here. It can be there, here and now, and it can show itself. This trace appears in, of all places, the written *appearance* of the signature, beyond the "face-to-face" encounter. It shows in the unqualifiable, non-repeatable specificity of *the particular traits* of *this* handwriting, which reveal themselves precisely in the moment that they withdraw from any sort of meaning. However, in order to do so, the signature must be *visible, it has to show itself.* Thus, there is an *aesthesis* that, for systematic reasons, *has to precede* all meanings and denotable forms, the withdrawals of writing, and even intentional consciousness. Before all that, there is something present that shows itself (Mersch 2002, 346). Derrida too fears the false return of intentional consciousness in such appearances. He fears being "seduced" "by the word *ductus*, by the necessity of its meaning," since it "signifies the idiomatic trait of the draftsman coming along to sign all by itself, before even the undersigning of the proper name. The *ductus* . . . is as good as a signature, so they say. . . . What, finally, *gives* [the] signature? What does it give? Is there (*gibt es?*) some signature?" (Derrida 1987, 193). Even when Derrida underscores the fact that "a mediation linking the question of secrecy to that of responsibility immediately raises the question of the name and of the signature" (Derrida 1995, 58), the seduction of the *ductus* harbors the danger of the sort of false address that motivates Levinas to write that an *aesthetic* epiphany is unthinkable. Derrida is not particularly stringent here either; indeed, he cannot be.

5. NOT EVEN A GHOST IN THE MACHINE

It is well-known that "Signature Event Context" ends by "giving" Derrida's name three times: in printed initials, as printed name, and as Xeroxed handwriting. These three versions stage the writing "I's" withdrawal into the differential, pluralizing iterability of writing that give "Derrida" as Other the chance to reside inaccessible in *différance*. Most of all, the Xeroxed signature demonstrates the very lack in which this iterability is grounded. The Xeroxed signature shows only one of many possible appearances of Derrida's signature, which each time bears singular, visible traits and as such will never have been the same twice. Because a person's handwriting can only enter into the calculus of exchange in the form of a *sign*. In the printed copy, the remainder of this exchange and the singularity that it sacrifices appears in the negative. But *they appear*, before the attribution of any meaning, as a "trace" of the singularity that is responsible for the specific traits of the signature: pen, ink, paper, desk, and whatever else cannot be controlled in the act of writing. Must not the withdrawal of writing somehow preserve itself in these traces and their erasure? Do not these traces, in a sense more Levinasian than Derridean, herald a presence in absence without meaning (and vice versa)? Does not this plurality of "traces" disseminate itself in multiple differences that leak through to the touch of a book's page and the appearance of its print? Does not Derrida's writing relish in playing with a certain *aesthesis* of language that undermines his definition of writing derived out of the formal structure of the sign?

Clearly, Derrida pluralizes the Other not only by deferring meaning, but also by playing with the sound and appearance of words. For instance, the inaudibility of the difference between *différence* and *différance* has to be made perceptible in the visual appearance of writing (Derrida 1982), giving *différance* a concealed presence that has to show itself in order to be registered at all (Mersch 2002, 346). This plural dimension of writing's differentiality—between sign, calculus of iteration, *aesthesis*, and materiality—make up a series of differences that bring a further, highly "im-possible" possibility of non-substitutability into play. But in the end, Derrida never explicitly addresses this dimension of writing. Deconstruction not only needs the works of others as challenges that "provoke us to think the event" (Derrida 2007, 360), but it also needs them as "exterior instances," *as things* "that preserve the asymmetrical relation to the Other" (Busch 2004, 9).

So, if Derrida's ethics is based on introducing the differentiality of writing into that impossible exchange called substitution, why does he exclude mediality and materiality from his discourse, even more so as he cannot avoid working with them? Given that substitution is the site where a response must take place, a response that cannot be articulated in any way *but* through mediation, it seems that substitution must be conceived of as a call for a mediation in which the medium of alterity stands in an uneasy relation with the alterity of the Other. This mediation must be given the impossible task of substituting for the Other, *without* being able to do so, *without* wanting to do so, and *without* any symmetry through a common humanity. Would that not be the very impossible gift of substitution? *Through* a mediality and materiality as the sites of an epiphany, as the sites of this "*via/dia*" (Mersch 2010)? Might an inconceivable Other be capable of befalling *via/through* the differential antagonism of mediality and materiality, via their transience and even via their abstract, lifeless otherness? But then, the spectrality of the machine would not only depend on the non-appearance of *différance*, but also on the appearance of this *via/dia*, which would pose a challenge to the *anaesthetics* of *différance*. What if this materiality were the only thing that could save writing from its horrifying, purely calculating zero level? Then there might be the possibility of a materially bound, self-revealing *aesthesis* that would not run the risk of reinstituting egology, and that would at once retain writing's resistance to formalism. The singularity of the trace of the Other would reside in the friction between the possibilities of infinite iterability and the materiality of the mediality that bears this iterability. Thus, on the one side, the singularity of the trace of the Other takes part in the singularity of specific traits, and on the other, it withdraws from every ascription of meaning, every codification, in showing itself. And neither moment can be attributed to an intentional consciousness, because both precede it. Does not this antagonism between multiple differences—including that between the human and "everything else"—bear within itself a truly dissymmetrical substitution without any sort of intentionality? Things and materialities that sometimes age and change have to substitute as bearers of the traces of the Other. This would place further limits on the formal iterability of immaterial signs. Things can hardly die, but they can pass away. And, in passing, they would take with them the traces of the Other.

It is precisely in this material dimension that a concept of writing can be differentiated from the formal language of computer programs. Materiality and embodiment resist an unlimited applicability of the program's formalism, which ultimately only ever feeds back into its own formal calculus. The purely formal systems of computer programs not only have the potential to get caught up in their own feedback loops, regardless of who is using them, their very structure guarantees the continuity of this endless cycle. It is a truly identical repetition rather than a pluralization. Even taking the hardware into account—the materiality of the machine—does not really help, as it only affects how fast the apparatus can carry out a certain quantity of work. This logic of the self-identical is underscored by the fact that backing-up a hard drive can mitigate the loss of data caused by a crash. It all revolves around data, regardless of the specific hard drive that stores it. User settings can always be restored. Contents can always be exchanged. The data might be unbound from origins and contexts, but as self-identical data, it is also free of all alterity. In the absence of any kind of *alter* in the iterability, the possibility of an event and of otherness is eliminated.

But maybe this re-reading of writing is itself a symptom of the calculated formalism of program writing, of a mere vestige of writing (Mersch 2006). Kittler's dictum that "both people and computers are . . . run by programs" (Kittler 1999, 17) might in retrospect prove itself to be a way of violently banishing the false humanism that both Derrida and Levinas have a tough time avoiding. Kittler's dictum also helps us see why digital violence takes on a life of its own in such a monological fashion, without any consideration of an Other. The purely formal relations of the formalisms running the network *precede* any possible relatedness to the Other. This is why some technologies of the self that use *Instagram* or communication via visual telephony might preserve a remainder that promises to hold onto the possibility of a multiple, differential trace of a trace of the Other. That is, users might seek in these practices a residual space where others still might appear exposed and naked, where the epiphany of the face of the Other might be possible, even in posting pictures of people jumping on *Instagram*. Because even without depicting the actual face of the person, photos still foster the minimal possibility that there is an inscription of a non-identical trace or an auratic *punctum* of the Other. But posting these pictures subjects them to the rules of exchange of a circulation-machine that inevitably recurs and returns to itself. Thus, there is only a relation to programs that pre-scribe every relation as a formal one (in the mathematical sense)—or as a relation to the apparatus itself. Nothing other stares out at us from the interfaces of social media. Apparatus and program, both are purely formalistic. They calculate, and are thus founded in circulation and return without responsivity. Power outages, crashes, pixilation, and noise are always just defects. There is not even a slight impossibility of an event of the Other's trace, because the program lacks in its very structure the *multiple* differences that would be able to open the digital *through an otherness towards an Other*, even if it only be a ghost in the machine.

NOTES

This chapter was translated by Adam Bresnahan.

1. At the same time, Kittler (1999, xxxix) fears that technologies "that not only subvert writing, but engulf it and carry it off along with so-called Man, render their own description impossible." That the pure calculus of mathematics is a diminishing of the concept of writing is underscored by Mersch (2006). On the various connections and contaminations between cybernetics and media theory, see Mersch 2013.

2. And yet, an ironic "distortion of comprehensibility" puts Derrida's readers in the position of (shared) responsibility; and Searle was not the only one to make this performativity of Derrida's writing misfire (Schumacher 2000, 259–337).

3. Though, Derrida, within his implicit humanism, also considers the vulnerability of the animal (Derrida 2008).

4. Levinas uses this to exclude portraits, for instance, as a legitimate means of encountering the face (Levinas 1999, 121–130).

REFERENCES

Bataille, Georges. 1988. *The Accursed Share: An Essay on General Economy, Vol. 1: Consumption.* New York: Zone Books.

Bernasconi, Robert. 2002. "What is the Question to Which 'Substitution' is the Answer?" In *The Cambridge Companion to Levinas*, edited by Simon Critchley and Robert Bernasconi, 234–251. New York: Cambridge University Press.

Bourdieu, Pierre. 1998. "The Economy of Social Goods." In idem, *Practical Reason: On the Theory of Action*, 92–123. Stanford: Stanford University Press.

Busch, Kathrin. 2004. *Geschicktes Geben: Aporien der Gabe bei Jacques Derrida*. Munich: Fink.

de Man, Paul. 1979. *Allegories of Reading: Figural Language in Rousseau, Nietzsche and Proust*. New Haven and London: Yale University Press.

Derrida, Jacques. 1978. "Violence and Metaphysics: An Essay on the Thought of Emmanuel Levinas." In *Writing and Difference*, translated by Alan Bass, 97–192. London and New York: Routledge.

Derrida, Jacques. 1982. "*Différance*." In *Margins of Philosophy*, 3–27. Chicago: University of Chicago Press.

Derrida, Jacques. 1987. *The Truth in Painting*, translated by Geoff Bennington and Ian McLeod. Chicago and London: The University of Chicago Press.

Derrida, Jacques. 1988a. "Signature Event Context," translated by Samuel Weber and Jeffrey Mehlman. In idem, *Limited Inc*, 1–24. Evanston, IL: Northwestern University Press.

Derrida, Jacques. 1988b. "Limited Inc a b c . . . ," translated by Samuel Weber. In idem, *Limited Inc*, 29–110. Evanston, IL: Northwestern University Press.

Derrida, Jacques. 1992. *Given Time: Counterfeit Money*, translated by Peggy Kamuf. Chicago and London: The University of Chicago Press.

Derrida, Jacques. 1995. *The Gift of Death*, translated by David Wills. Chicago and London: The University of Chicago Press.

Derrida, Jacques. 2001. "The Deaths of Roland Barthes." In idem, *The Work of Mourning*, edited and translated by Pascale-Anne Brault and Michael Naas. Chicago and London: The University of Chicago Press.

Derrida, Jacques. 2002. "Typewriter Ribbon: Limited Ink (2)." In idem, *Without Alibi*, edited and translated by Peggy Kamuf, 71–160. Stanford: Stanford University Press.

Derrida, Jacques. 2007. "My Chances/*Mes chances*: A Rendezvous with Some Epicurean Stereophonies," translated by Irene Harvey and Avital Ronell. In idem, *Psyche: Inventions of the Other, Volume 1*, 344–376. Stanford: Stanford University Press.

Derrida, Jacques. 2008. *The Animal That Therefore I Am*, edited by Marie-Louise Mallet, translated by David Wills. New York: Fordham University Press.

García Düttmann, Alexander. 2008. *Derrida und ich: Das Problem der Dekonstruktion*. Bielefeld: transcript.

Han, Byung-Chul. 2016. "Der springende Mensch," *Die Zeit*, no. 4, 21 January 2016.

Heidegger, Martin. 1977. *Sein und Zeit: Heidegger Gesamtausgabe* 2, Frankfurt/Main: Klostermann.

Hörl, Erich. 2005. *Die heiligen Kanäle: Über die archaische Illusion der Kommunikation*. Zurich: Diaphanes.

Kittler, Friedrich A. 1980. "Einleitung." In *Austreibung des Geistes aus den Geisteswissenschaften. Programm des Poststrukturalismus*, edited by. Friedrich A. Kittler, 7–14. Paderborn, Munich, Vienna and Zurich: Schöningh.

Kittler, Friedrich A. 1999. *Gramophone, Film, Typewriter*, translated by Geoffrey Winthrop-Young and Michael Wutz. Stanford: Stanford University Press.

Levinas, Emmanuel. 1986. "The Trace of the Other." In *Deconstruction in Context. Literature and Philosophy*, edited by Mark C. Taylor, 345–359. Chicago & London: The University of Chicago Press.

Levinas, Emmanuel. 1987. "Reality and its Shadow." In *Collected Philosophical Papers*, Pittsburgh: Duquesne University Press.

Levinas, Emmanuel. 1996. "Jacques Derrida: Wholly Otherwise." In idem, *Proper Names*, translated by Michael B. Smith, 55–62. Stanford: Stanford University Press.

Levinas, Emmanuel. 1998. *Otherwise Than Being, or, Beyond Essence*, translated by Alphonso Lingis. Pittsburgh: Duquesne University Press.

Levinas, Emmanuel. 1999. *Alterity and Transcendence*, translated by Michael B. Smith. London: The Athlone Press.

Mauss, Marcel. 1966. *The Gift. Forms and Functions of Exchange in Archaic Societies*. London: Cohen & West.

Mersch, Dieter. 2002. *Was sich zeigt: Materialität Präsenz Ereignis*. Munich: Fink.

Mersch, Dieter. 2005. "Die Geburt der Mathematik aus der Struktur der Schrift." In *Schrift: Kulturtechnik zwischen Auge, Hand und Maschine*, edited by Gernot Grube, Werner Kogge and Sybille Krämer, 211–233. Munich: Fink.

Mersch, Dieter. 2006. *Medientheorien zur Einführung*. Hamburg: Junius.

Mersch, Dieter. 2010. "*Meta/Dia:* Zwei unterschiedliche Zugänge zum Medialen." In *Zeitschrift für Medien- und Kulturforschung* 2, 185–208.

Mersch, Dieter. 2013. *Ordo ab chao – Order from Noise*, Zurich: diaphanes.

Schumacher, Eckhard. 2000. *Die Ironie der Unverständlichkeit. Johann Georg Hamann, Friedrich Schlegel, Jacques Derrida, Paul de Man*, Frankfurt/Main: Suhrkamp.

Weber, Sara. 2015. "Geschlossene Gesellschaft: Diese Frau will Facebook überflüssig machen," *Süddeutsche Zeitung*, 30 November 2015, http://www.sueddeutsche.de/digital/soziale-netzwerke-diese-frau-will-facebook-ueberfluessig-machen-1.2801813?reduced=true.

Wirth, Uwe. 2007. "Zwischen genuiner und degenerierter Indexikalität: Eine Peircesche Perspektive auf Derridas und Freuds Spurbegriff." In *Spur: Spurenlesen als Orientierungstechnik und Wissenskunst*, edited by Sybille Krämer, Werner Kogge, and Gernot Grube, 55–81. Frankfurt/Main: Suhrkamp.

Chapter 3

Performative Modalities of Otherness

Jörg Sternagel

DYNAMIC CORRELATIONS

The term *performative* has been used very differently in linguistics, cultural studies, literature, theater, dance, and media studies as well as in philosophy. While it has been applied in manifold ways to scrutinize speech, theatrical or ritual acts, and procedures of writing and reading, all the various approaches share an interest in the conditions of cultural and social contexts which emerge from events and practices. *Concepts of the performative* place our attention *on* and *in* the world.[1] In so far as cultural and social transformations derive from events and practices, our human reality is not simply a given. It cannot be grasped by facts and numbers alone, but must rather be acknowledged through reference to dynamic correlations in space and time, to dimensions that are describable in terms of activity and creation. For, we do not only speak of the world, but, in speaking, act in and on the world; and we are not only in the world, on our own, as autonomous subjects so to speak, but we exist with others, with other human beings. We act and react with our language, our voices, our faces, our bodies and gestures. This gives rise to a host of ethical dimensions of the performative which unfold and enfold with regard to *(non-)speaking and (non-)acting, being and alterity*, as well as *mediality and mise-en-scène*.

In cultural processes, these ethical dimensions anchor language, voice, body, and gesture in social and cultural reality, and make visible and palpable the effects and correlations thereof. Primordial as they are, ethical dimensions touch upon the realms of human activity and inactivity, acting and non-acting, speaking and non-speaking, success and failure. In an open field between language, the constitution of gender subjectivity and embodied practices, between productive and unproductive phenomena, these ethical planes serve as trajectories for *performative modalities of otherness*, linking performative theory with philosophical reflection on the ethics of human conduct, the human body, the other, perception, and things. Thus, whilet reflection on *(non-) speaking and (non-)acting* considers ways to describe the world as it is and how it works, questions pertaining to *being and alterity* provoke doubt with regard to the secure place of the subject as it is continuously questioned by the other. Investigations into the *mediality and mise-en-scène* of cultural performances locate language, the voice, face, body, and gesture in interrelations where effects are generated, and words become deeds.[2]

How do we place ourselves in the world, and relate to other human beings? Does being necessarily mean *being acknowledged* by others? And is this reference to others necessarily exclusively human? Or can it be broadened to include other beings, such as animals and things? How can these relationships be described? Does an ontological framework suffice, or is an ethical terminology needed that reaches beyond phenomenology, especially beyond intentional analysis? How, and why, can the body, and in particular, a person's face, constitute a resistant claim? What are their respective performative *effects*?

ETHICAL DIMENSIONS OF THE PERFORMATIVE

Within correlations between being and alterity, sense and meaning are situated in the ethical and the performative. Uniting the subjectivity of *perceiving*—intentionally aiming at an object—with the objectivity of *expressing*, *creating*, such as language, a play on stage or in film, or a dance, reflections begin on an ontological order of corporeal, linguistic, and artistic gestures, where the encounter with another is in a cultural whole, in which she or he is present and illuminated by this whole, as a text is by its context. Drawing inspiration particularly from the essay "Meaning and Sense" by Emmanuel Levinas, this order is gradually questioned, and the encounter with the other, her or his corporeal, linguistic, and artistic gestures, turns to an epiphany that involves a signifying of its own, dependent on sense, but independent of meaning: I am under a basic obligation, the other reveals itself as a command, as an ethical and performative imperative, *before culture and aesthetics* (Levinas 1996a). Engaging in a *conversation* by drawing attention to correlations between *language, poetry,* and *alterity*, meaning and sense are situated in the ethical and the performative. Our conversation as such, in its unfolding reflection, correspondingly might point to the action of living or having one's being in a place or among persons, it might even, as one entry in the Oxford English Dictionary suggests, point *to actions of consorting or having dealings with others, living together, commerce, intercourse, society, as well as intimacy*. This conversation begins with a twist of the *genetivus subjectivus* and the *genetivus objectivus*—in other words, in a movement from the "I," the subject, her or his secure and second place *toward* the Other, the other human being and her or his demand, from the *first* and insecure place. From this first place, our conversation is joined by Levinas who adds that we are not the subject of the world and a part of the world from two different points of view; moreover, *in expression*, we are subject *and* part of the world at once—"to perceive is both to receive and to express," "we know through gestures how to imitate the visible and to coincide *kinesthetically* with the gesture *seen*: in perception *our* body is also the 'delegate,' the representative of *Being*" (Levinas 1996a, 41).

Levinas offers a phenomenological perspective on *experience* and *language*, while especially defending phenomenology against a positivistic form of empiricism that reduces meaning in language, also in metaphorical and literal variation, to contents given to consciousness. *There is no given already possessing identity; no given could enter thought simply through a shock against the wall of receptivity*. Language in particular, and the overall experience *signify* on the basis of the world and of the position of the one that looks at them—in uniting the subjectivity of perceiving—intentionally aiming at an object—with the objective of expressing, an operation that creates, such as language on an ontological order of corporeal, linguistic, and artistic gestures, where the encounter with another is in a cultural whole, in the place of the *horizon*, in which he or she is present and illuminated by this whole, as a text is by its context. In *expression*, we are subject and part at once. Throughout this phenomenological conception, expression defines culture, whereas culture and artistic creation are part of the ontological order themselves, or rather, they are ontological "par excellence," since they make the understanding of Being possible. In his discussion of this conception, Levinas remains especially interested in language, more specifically in *language qua expression (in the capacity of expression, as being expression)* wherein, above all, he locates the creative language of poetry. Levinas though is not solely reflecting on phenomenology for the reason of defending its approach, but also for the reason of going beyond its cultural, social, symbolic and linguistic implications and significations, for gradually questioning them—since Levinas' overall project is to escape *being*, whereas *escape*, as he suggests in his first thematic essay *On Escape* is the need to get out of oneself, that is, to break that most radical and unalterably binding of chains, the fact that the I is oneself (Levinas 2003).

Levinas' philosophy moves beyond the horizon of Being, toward the Other, which is also the reason why Levinas prefers to describe meaning less in terms of signification, but more with the movement of sense, a unique sense that is imperative, that is revealed in the emergence of the Other, in the other human being as the first place I encounter in the world, while this encounter with the Other, her or his corporeal, linguistic, and artistic gestures, turns to an epiphany, an immediate manifestation, making an entrance, dependent on sense, but independent of meaning. While culture and artistic creation "make the understanding of being possible," *language qua expression*, and above all, the creative language of poetry, move toward the other, or escapes being (Levinas 1996a, 41). Levinas explicitly refers to the poetry of Paul Celan, the German language poet of Romanian Jewish origin, who shows us that "poetry is ahead of us," as in a speech titled *The Meridian* on the occasion of receiving the Georg Büchner prize in Darmstadt, West Germany in 1960, where he meditates on poetry. Celan says, "The poem hold its ground on its own margin. In order to endure, it constantly calls and pulls itself back from an 'already-no-more' into a 'still-here.' The poem is lonely. It is lonely and *en route*. Its author stays with it. Does this very fact not place the poem already here, at its inception, in the encounter, in the *mystery of encounter*? The poem intends another, needs this other, needs an opposite. It goes toward it, bespeaks it" (Celan 2003, 49). The poem is situated at the pre-syntactic and pre-logical level as well as at the pre-disclosing level: "at the moment of pure touching, pure contact, grasping, squeezing," which is, for Levinas, also a "way of giving, right up to and including the hand that gives" (Levinas 1996b, 41).

I see no essential difference between a handshake and a poem. It is a "language of proximity for proximity's sake," it is the first of the languages, response preceding the question, responsibility for the neighbor, by its *for the other*, the whole marvel of giving (Levinas 1996b, 41). Here, Levinas' notions of proximity, responsiveness, and responsibility are *key*: proximity, however, should not be confused with affinity, where two bodies are close to each other, corresponding to their positions and the intersects they share, where remoteness and nearness are already always relative. Levinas' understanding of proximity originates from *the* otherness, in a movement from mimesis to alterity, from being to the other, from beginning oneself to beginning elsewhere, in re-defining freedom, whereas the other desires without wanting. While Levinas develops his ethics along with topics like time, space, body, and the senses, again in discussions of phenomenological findings, he goes beyond them: *The neighbour in Levinas is more a stranger than a friend.* The other reveals herself or himself as a command, as an ethical imperative, in an event which arrives, but not as if I who receive the command am already somebody before responding to the command. The event of this command is neither a responsible act nor a neutral fact, moreover, it implies responsiveness and responsibility, provoking our response of responsibility. I am under a basic obligation in my own sphere, from which I am confronted and respond to what is not mine, what interferes, what is owned by the other, what is demanded by the face of the other, what infinitely exceeds the limits of any order. This inflection appears, *while* I am responding to it, in between the other and myself. It is marked by asymmetry. It neither follows a scheme of stimulus and reaction nor a principle of cause and effect, as the effect precedes the cause and points to a gap of impossibility as soon as my own sense-making and sensual possibilities are overcome. We arrive at the movement from the "I," the subject, her or his secure and second place *toward* the Other, the other human being and her or his demand, from the *first* and insecure place, at a *beginning oneself from beginning elsewhere*, and at the poem which goes to the other and hopes to find her or him freed and vacant, in the circularity of this movement, this trajectory, this meridian, where the poem becomes conversation, addressing, questioning, *performing*, pointing toward open, empty, free spaces, preceding all thematization, letting the otherness's ownmost also speak: the time of the other whose demand precedes our response and separates it with its temporal delay, leaving the real its alterity, seeking to inhabit the in-between. As the poem, that "intends another, needs this other, needs an opposite, goes toward it, bespeaks it." For the poem, everything and everybody is a figure of this other toward which it is heading. For the words of the poem, their power to affect in their temporality at the event of speaking them, gives a priority of sound over semantics. Their imperative sound means the presence of others making themselves affective in advance of what is said. Their expression unfolds an ethical event as in Paul Celan's poem "I Can Still See You." "Your face quietly shies," it reads, while every moment of memory speaks a painful "never." The poem is about love, from a loved one, to the beloved, where love is the unique situation in which the Other appears, as an object of love, that is an object of need—but at the same time the Other remains totally other, the pure transcendence (Celan 1972, 99). In love I am called to responsibility because the beloved one appears to me in all its fragility and weakness; in love, I fear for the Other.

THE FACTICITY OF MEN AND THE WORLD

In the preface of his *Phenomenology of Perception*, Maurice Merleau-Ponty elaborates on the philosophy of phenomenology, highlighting the example of perception, and showing *how* the body plays crucial roles in perception, speech, sexuality, and our relation to others. Merleau-Ponty defines phenomenology as the study of essences, "according to it, all problems amount to finding definitions of essences: the essence of perception, or the essence of consciousness, for example" (2002, xii). He continues by pointing out that these "essences" are also brought back "into existence" by the philosophy of phenomenology, while not expecting "to arrive at an understanding of man and the world from any starting point other than that of their 'facticity'" (Merleau-Ponty 2002, xii). It is this *facticity* that serves as a starting point for a study of the essence of experience, and also as a starting point of our experience of technics and media; it allows a direct description of our experience as it is, without, as Merleau-Ponty (2002, xii) suggests, "taking account of its psychological origin and the causal explanations which the scientist, the historian or the sociologist may be able to provide." Hence, phenomenology serves as a close descriptive method that, as David Kaplan (2004, 91) remarks, "is based on the idea that experience is always relational," and that "every instance of experience has its reference or direction toward what is experienced." The aim of this method, Kaplan continues, is "to identify the essential or invariant features of experienced phenomena."

Two of these experienced phenomena are *technics* and *media*, and the phenomenological description of both of them aims at an understanding of how they affect us, based on the conviction that we cannot understand them unless we realize them as objects in relation to us, and not merely as objects without a relation to us, or, more directly, as merely objects in themselves. The description of these experienced phenomena is always a description drawn from my own knowledge of the world as it is there before any possible analysis of mine. It is, as Merleau-Ponty (2002, xii) notes, "not an object such that I have in my possession the law of its making; it is the natural setting of, and field for, all my thoughts and all my explicit perceptions." Starting from the natural setting of, and the field for all my thoughts and all my explicit perceptions, it is important to identify phenomenology as the study of human experience, and of the ways things present themselves, of the ways they, as Daniel Frampton (2006, 40) elaborates, "present themselves to us in and through such experience. Phenomenology deals with 'appearances', asserting that appearances are real, that they belong to being." It is a phenomenology that sees subject and object as inseparable, it is a phenomenology that is existential, because it deals with the embodied nature of human consciousness, and is therefore not transcendental, because it does not deal with universal descriptions, it rather prefers an embodied description. We, as human beings, with our lived bodies, as spectators or players, experience, for example, the visual and sound material of a film or video game, and address ourselves to it by seeing, hearing, and moving; we direct our intention toward it while gazing at it, we perceive, we are an active part of the experience with this material.

We perceive the material as sensible, as making sense; we see and think the objects before us; we take them; we make them our own; they are lived in experience, and our lived bodies are, as Vivian Sobchack (1992, 40) suggests, "both agent and agency of an engagement with the world that is lived in its *subjective* modality as *perception* and in its *objective* modality as *expression*, both modes constituting the *unity* of meaningful experience." Every perception is a communication of our bodies with things, of our lived bodies and the perceived world. In the lengthy chapter on the experience of space in his *Phenomenology of Perception*, Merleau-Ponty assumes a precise correlation between the lived body and the perceived world, while discussing an experiment of the gestalt psychologist Max Wertheimer: "What counts for the orientation of the spectacle is not my body as it in fact is, as a thing in objective space, but as a system of possible actions, a virtual body with its 'phenomenal' place defined by its task and situation. My body is wherever there is something to be done" (Merleau-Ponty 2002, 291). Indeed, the body of the protagonist of Wertheimer's experiment is wherever there is something to be done: The psychologist places the man in a room, where he can only see the room through a mirror that reflects it in an angle at forty-five degrees to the vertical, leading to the first experience of seeing the room "slantwise," of seeing another man walking in this room leaning to one side while going, of having the impression of a piece of cardboard falling down the door-frame "to be falling obliquely," of having a general "queer" effect. But, after a few minutes, the protagonist's experience changes: "the walls, the man walking about the room, and the line in which the cardboard falls become vertical." "It is as if certain objects (walls, doors and the body of the man in the room), having been seen aslant in relation to a given level, then take it upon themselves the vertical, acting as 'anchoring points', and causing the previously established horizontal to tilt sideways" (Merleau-Ponty 2002, 289).

What the philosopher is pointing at, with the support of the psychologist's experiment, is, as quoted earlier, that the body is "wherever there is something to be done," where action and perception point to "a perceptual ground, a basis of my life, a general setting in which my body can co-exist with the world" (Merleau-Ponty 2002, 290). Bodily existence is therefore an existence that is oriented to action, it is an existence that correlates with the world surrounding it, it is open to action in the world. The body's "coexistence with the world magnetizes experience and induces a direction in it" (Merleau-Ponty 2002, 293). For a phenomenology of technics and media, as for all other phenomenologies, it is exactly this correlation that serves as the basic understanding of the body, my body, and the world, the world surrounding me. It is a sensitivity for this correlation that Merleau-Ponty elaborates when discussing the role of technologies within the realm of perception and daily life, within the correlations of the lived body, or its experienced spatiality, and artifacts. Merleau-Ponty regards the lived body as an experienced bodily spatiality which can be extendible through artifacts. In the chapter on the spatiality of one's own body and motility in *Phenomenology of Perception*, a feather in a woman's hat serves as an example for the body's comprehension of movement and spatiality: "A woman may, without any calculation, keep a safe distance between the feather in her hat and things which might break it off. She feels where the feather is just as we feel where our hand is" (Merleau-Ponty 2002, 165). Perception can be materially extended through the body of an artifact, and is not limited to my own body, to the surface of my skin. Merleau-Ponty substantiates his point with another example:

> The blind man's stick has ceased to be an object for him, and is no longer perceived for itself; its point has become an area of sensitivity, extending the scope and active radius of touch, and providing a parallel to sight. In the exploration of things, the length of the stick does not enter expressly as a middle term: the blind man is rather aware of it through the position of objects than of the position of objects through it. The position of things is immediately given through the extent of the reach which carries him to it, which compromises besides the arm's own reach the stick's range of action. If I want to get used to a stick, I try it by touching a few things with it, and eventually I have it 'well in hand', I can see what things are 'within reach' or out of reach of my stick. (Merleau-Ponty 2002, 165–166)

What is striking here is the basis for perception at a distance, which is mediated through the artifact, the instrument, the stick, the technology. To get used to the stick, for example, and to other technologies is "to be transplanted into them, or conversely, to incorporate them into the bulk" of my own body. The author offers another example, and directs his phenomenology of perception, and of instruments, to the question of habit: "It is possible to know how to type without being able to say where the letters which make the words are to be found on the bank of keys. To know how to type is not, then, to know the place of each letter among the keys, nor even to have acquired a conditioned reflex for each one, which is set in motion by the letter as it comes before our eye" (Merleau-Ponty 2002, 166). To get used to a feather in a hat at safe distance, to a stick through the extent of the reach, and to a typewriter and the letters on the bank of keys are habits that express the power of "dilating my-being-in-the-world," or changing my existence by "appropriating fresh instruments" (Merleau-Ponty 2002, 166). We realize these technologies as objects in relation to us, and not merely as objects without a relation to us or, more directly, as merely objects in themselves. This relation is crucial for the understanding of active relational pairs.

TECHNICS AND MEDIA

A phenomenology of technics and media is grounded on its attention to detail and its emphasis on the relationship between the human being and technology, the human-technology relations. Its task is to discover the various structural features of those relations, its material, tactile, and tangible dimensions, and its impact on our perception. Its beginning is to discover the multiple ways in which my body interacts with the world surrounding it, with its environment, by means of technologies. For achieving such an intended result, Don Ihde (1990, 72) suggests to "focus upon experientially recognizable features that are centered upon the ways we are bodily engaged with technologies." While dealing with these ways of bodily engagement, he realizes three sets of existential technological relations with the world: (1) *embodiment*, (2) *hermeneutic*, and (3) *alterity* relations. Ihde begins by saying: "To embody one's praxis through technologies is ultimately an existential relation with the world." He continues: "I call this first set of existential technological relations with the world embodiment relations, because in this use context I take the technologies into my experiencing in a particular way by way of perceiving through such technologies and through the reflexive transformation of my perceptual and body sense" (Ihde 1990, 72). Ihde questions the relation between the technology and the body by asking where and how the technology is experienced. He discovers a doubly ambiguous positioning: first, the technology, in his words, "must be technically capable of being seen through, it must be transparent," it must be discovered (Ihde 1990, 73). Here, by using the term *technical*, he refers to the physical characteristics of the technology. Second, the technology as transparent becomes a possibility for *embodiment*. By relating to the material conditions for embodiment, he also refers to the activity of embodiment. Inspired by Merleau-Ponty and his study with the feather, the cane, and the typewriter, Ihde works with examples like a telescope or glasses. Both philosophers take account of the ways in which technologies may be embodied, elaborate on the material extension of perception by technology, and turn to the question of habit while using technology, while incorporating it into bodily action, while becoming part of the perceptual and bodily experience.

The central argument of Ihde's focus on embodied technics is his understanding of technics as "the symbiosis of artifact and user within a human action," pointing to embodiment relations that are not restricted to visual relations, and that "may occur for any sensory or microperceptual dimension" (Ihde 1990, 73). Within these relations, the body, or the embodied subject, my body, responds to appeals made by these embodied technics to all five senses: sight, sound, smell, taste, and touch. The full bodily awareness is part of the ordinary experience of the world of artifact and user. My body gets used to the artifact, is extended by its capacities, and is also transformed by it. My body experiences something else through technology; it has mediated access to his lifeworld through technology. "Embodiment relations constitute one existential form of the full range of the human-technology field" (Ihde 1990, 80).

Dealing with bodily engagement, Ihde realizes a second relation between the body and technology: opposed to the embodiment relation, he sees a *hermeneutic* relation, in which the technology is also experienced as an artifact with which my body experiences something else, but here my body does not experience the world surrounding it, my lifeworld. It experiences a text; a text in the sense of a representational artifact. Ihde (1990, 80) explains:

> The term hermeneutic has a long history. In its broadest and simplest sense it means "interpretation," but in a more specialized sense it refers to textual interpretation and thus entails reading. I shall retain both these senses and take hermeneutic to mean a special interpretive action within the technological context. That kind of activity calls for special modes of action and perception, modes analogous to the reading process.

Ihde directs attention to a different mode of perception. The perception of technology in a hermeneutic relation is an interpretive act, as in the act of reading, of reading a writing, a text, that presents itself as the "world" of the text. He discovers a presence that is hermeneutic: the world is experienced as represented and referred to by my body. It is not a lifeworld that is immediately present for my body as in the embodiment relation. "The world is linguistically mediated, and while the words may elicit all sorts of imaginative and perceptual phenomena, it is through language that such phenomena occur" (Ihde 1990,

84). My body experiences another extension through interpretive and linguistic capacities in the hermeneutic relation. This extension adds to the extension of sensory capacities in the embodiment relation, but within a movement from the first to the second relation along a "human-technology continuum."

Moreover, for Ihde, a third relation of body and technology exists beyond the relations he calls embodied and hermeneutic. He defines the third one as the alterity relation. Ihde continues with a question: *In what phenomenological senses can a technology be other?* While in the first two relations, my body experiences extensions and transformations through technology, my body is now experiencing the technology as an artifact that is personalized, within the range of anthropomorphism, according to Ihde (1990, 98), "reaching from serious artifact-human analogues to trivial and harmless affections for artifacts." The most striking example of this relation, that also implies the relations of embodiment and of the hermeneutic kind, is my body's engagement with quasi-otherness and my body's experience of quasi-otherness while playing video games:

> In the actual use of video games, of course, the embodiment and hermeneutic relational dimensions are present. The joystick that embodies hand and eye coordination skills extends the player into the displayed field. The field itself displays some hermeneutic context (usually either some "invader" mini-world or some sports analogue), but this context does not refer beyond itself into a worldly reference. (Ihde 1990, 100)

In addition to these dimensions, Ihde emphasizes, there is the dimension of the sense of my body interacting with "something other" than my body. There is the sense of "interacting with the technological competitor" in a kind of "dialogue or exchange." There is the fascination of and challenge with the quasi-otherness, with the animation on screen. My body experiences the animation as something different; my body interacts with *the* something other. I phenomenologically understand the technologies as active relational pairs, as pairs of human-technology. I realize the ways of bodily engagement within the sets of existential technological relations with the world.

The description of the experienced phenomena is always a description drawn from my own knowledge of the world as it is there before any possible analysis of mine. The descriptive method is relational, and every moment of experience, as Kaplan (2004, 91) stresses, has "its reference or direction to what is experienced." The outcome of this method is the realization of technologies as means of transformation one's sense of one's body, of transformation of the bodily experience, based on interaction, in the act of experiencing. Going back to the room from above, as experienced through a mirror, remember Merleau-Ponty's description of the body, of my body as "a system of possible actions, a virtual body with its 'phenomenal' place defined by its task and situation," that "my body is wherever there is something to be done," where action and perception point to "a perceptual ground, a basis of my life, a general setting in which my body can co-exist with the world," making my body an active part of experience with media, with audiovisual media, and its perception and expression of time, space, and subjectivity, its perception of the visual and the visible (Merleau-Ponty 2002, 291).

BODIES, SCREENS, RENDERINGS

Investigations of ethical dimensions of the performative which unfold and enfold with regard to the *mediality and mise-en-scène* of *digital* performances focus on renderings as mediation between bodies and screens, *offstage*, where the manifestation of the human form gradually becomes invisible, while its rendering successively becomes visible. The perception of the visual and the visible while playing a video game, where interactivity and simulation lead to bodily action, provoking synesthetic effects, and adding tactile and kinetic aspects to the experience, can be described as follows: My body is extended to the displayed field. My body experiences the animation as something other. My body interacts with this something other, with the avatar in an avatar-based single-player computer game as one possibility. The avatar is both an extension and a model. It mediates fictional agency. It is, as Rune Klevjer (2006, 89) suggests, "an instrument and mechanism, that defines a fictional body for the participant" and is therefore "an embodied incarnation of the acting subject," "dependent on the principle of the model." Understood as a model, the avatar, like the avatar "Mario" in Nintendo's *Super Mario Bros.* game series, defines a space of possibilities for the player and his fictional agency within the game. The avatar's objective properties are based on the capabilities and restrictions of the model, leading to the definition of "the boundaries of embodied make-believe" (Klevjer 2006, 89). The avatar's game, Mario's game, with its phenomenology of the body, appeals to my body through the game, through Mario. It is, quoting Klevjer (2006, 95), "the mediation of embodied agency that makes us relate to the avatar intuitively as an 'I can'"; the avatar "transforms bodily space, it transforms the space of potential action for the 'I can,' and it integrates with the body as a perceptual habit." My body experiences the avatar as something other, a virtual body, and actively interacts with it, while simultaneously inhabiting the simulated environment of the game, while simultaneously being implicated. Unlike the tool, unlike the cane and the typewriter, understood as instrumental extensions, the avatar is defined as a reflexive extension. It is, as Klevjer (2006, 98) concludes, "not just acting upon, but also being acted upon and affected by"; it is a body that exposes itself.

Especially learning from theories of computer games, actors, for example, gradually lose their background, their bodies, in front of the blue or green screen, surrendering their descent (sankr. *avatara*), down (*ava*) to fragments, passing over (*tar*) any human relation to reconsiderations and restructurings in digital codes. The spectator watches a digital performance and becomes involved with *another, the "alienness" of an avatar,* its opaque impenetrability that does not establish a relation. Instabilities now appear, where hybridization generates indecidabilities, where the digital programming of avatars is tied to a misguided realism and we, *without consideration,* do what we wish with it—going as far as a distortion of all familiar characteristics of *humanitas*. In this case, the experience of self and other and the manifold threads tying together performativity and responsivity systematically come apart and are *destabilized* (Sternagel, Levitt, & Mersch 2012, 51–58). From the ubiquity of screens to a wide variety of features related to media convergence, spectatorship and the modes of performance are undergoing a radical transformation. Media forms play across multiple platforms, and new genres—particularly ones that revise and reconfigure forms of "reality," on the one hand, or develop extended means for the depiction of perceptually realistic fantasy, on the other—emerge from these new conditions. One key site of transformation is the shift from celluloid to digital. Many films today have little relation to classical film and its materiality, that is, celluloid. They are shot, edited, and projected digitally, and even films that are shot on film-material today typically pass through digital phases in editing or projection. The mutability of the digital image, the extreme ease with which it may be manipulated, is engendering new types of performance and new challenges to its theorization.

In making an overall sense of the experience of film, of bodily activity and of perception, in linking image, mind and body, while meeting in close interrelation and emerging in their conjunction, this phenomenological description is also supported by Merleau-Ponty and considered in his lecture "The Film and the New Psychology": "Films directly present to us that special way of being in the world, of dealing with things and other people, which we see in the sign language of gesture and gaze and which clearly defines each person we know" (Merleau-Ponty 1964, 58). The affective embodied experiences with film are limitless and stress the materiality of film: "The meaning of a film is incorporated into its rhythm just as the meaning of a gesture may immediately be read in that gesture: the film does not mean anything but itself. . . . A movie is not thought; it is perceived" (Merleau-Ponty 1964, 57–58). Film offers the conduct or behavior of man to us, the spectators; it gives conduct or behavior as gripping possibilities. Film is the presentation of the spectator's being-in-the-world. It directly provides ways of behaving, and visibly demonstrates the organization and understanding of the perceptual field. Meaning in film is experienced, and originates in the mutual exchange of the personal intentions of the viewer and the movements of others, including the actors on the screen. The absorption in the film experience is a form of our mutual absorption in the world. Watching film, Merleau-Ponty (1964, 58) stresses, that "the film is not a sum total of images, but a temporal gestalt," the film is a paradigm of our inherence of the self in the world and in others, a paradigm of our relation to the world of objects and others; "movies are peculiarly suited to make manifest the union of mind and body, mind and world, and the expression of one in the other."

SPECTACLE, RECOGNITION, BEING

Merleau-Ponty (1964, 54) elaborates on the films' material to describe "the mingling of consciousness with the world, its involvement in a body, and its coexistence with others." We experience material, and address ourselves to it by seeing, hearing, and moving. We direct our intention toward it while gazing at it. We perceive. We are an active part of the film experience. We perceive film as sensible, as making sense, also digital film, film designed with computer-generated imagery, during distorted experiences with digital vicissitudes of an actor's face, for example, in Tim Burton's version of *Alice in Wonderland* from 2010, as reviewed by Al Hoff in the *Pittsburgh City Paper*:

> I did wonder whether Burton has exhausted his bag of tricks. Preening Johnny Depp in a fright wig? Lady-love Helena Bonham Carter in supporting role? Colorful but off-kilter sets? Danny Elfman score? Check, check, check, check. I could have forgiven it all if this film hadn't been so boring. Yet despite appending bookends about Victorian garden parties and international commerce, and making Alice (Mia Wasikowska) a full-grown woman on a return visit to the "wonderland" of her youthful dreams, a coherent, compelling storyline never emerges. Alice encounters all of the requisite creatures from Carroll's books (*Alice*, *Through the Looking Glass*), such as the Cheshire Cat, the Mad Hatter (Depp), the Red Queen (Carter), Tweedledum and Tweedledee, and the White Rabbit. But now there are chase scenes, magic swords, inter-kingdom warfare, a dragon (!) to be vanquished, and some holiday that makes the Mad Hatter channel Michael Jackson. It's alternately dull and frantic, with the pretty but vapid Wasikowska and the madcap Depp quickly wearing out their welcomes. The bright spot in the film is Carter, whom Burton depicts as a gigantic Elizabeth the first-type head atop a much smaller body. Carter delivers her lines with a hearty, demented brio that isn't as self-consciously showy as Depp's endlessly morphing accents and facial tics. I also adored all the heart motifs that the production designer worked into the queen's stylings. But viewers searching for the more elusive heart—the sad-sweet centers that mark Burton's better films—will be disappointed. There's just a lot of meaningless, color-saturated clutter and clatter down this rabbit hole.

In the colorful light of Burton's post-converted three-dimensional film, three issues of Hoff's review are striking: *First*, the public and private relation between Helena Bonham Carter and the film's director he shows interest in. *Second*, the film's clutter and clatter he addresses. And *third*, the only bright spot that developed in his experience: the depiction and performance of actress Bonham Carter. These three issues provide three paradigms of visibility.[3] The first paradigm, dealing with the celebrity Bonham Carter, will be presented as *the visibility of the spectacle*. The second, in the context of the film's clutter and clatter, is the *visibility of recognition.* And the third, starting with the reviewer's experience, within contexts of affect, pathos and the performative, is the *visibility of being*.

Visibility of the Spectacle

Before the film, Helena Bonham Carter, the actress works up interests in the work, in her fellow-actors, in props and set designs, in the camera, to bring, at least in part, the Red Queen to life. In sum, her interest, at least on an artistic level, would be to create a role. Before the film, I, in turn, work up an interest for Helena Bonham Carter, as I have experienced her in numerous films previously, as in *Howard's End* (James Ivory, 1992) and *Fight Club* (David Fincher, 1999), or in *Planet of the Apes* (2001) and *Sweeney Todd* (2007), also directed by Tim Burton, and therefore decide to be at the movies with her, to watch her in another film. In sum, my interest, on an experiential level, would be to meet again. During the film, Bonham Carter offers variations of her art, variations I am partly familiar with, like the tone of her voice, her pronunciation of words, her facial expressions. Here, the focus is on the situation of perception during the film, a situation that is also influenced by experiences before or after the film. This is where the visibility of the spectacle develops, pointing successively lesser and lesser to Bonham Carter's familiar performing choices and my experience of them, but more and more to her star or celebrity functions. Such a paradigm could be further scrutinized with Richard Dyer, for example, and his thoughts about *Heavenly Bodies*, where the economic importance comes into play and the star is also presented as being fashioned "out of the raw material of the person," where "make-up, coiffure, clothing, dieting and body-building can all make more or less of the body features they start with" (Dyer 1986, 5). Or, it could be interpreted with Guy Debord and his approach in the *Society of the Spectacle* (1962), where I am separated from the celebrity or star in everyday life as the images I see with her, from her are detached from life and only propose an illusionary form of life's unity, where Bonham Carter is "the spectacular representation of a living human being, embodying this banality by embodying the image of a possible role" (Debord 1995, §60). At this point, to move to the next paradigm, a further reflection upon the film's clutter and clatter Hoff addresses as well as on Bonham Carter's position in this mess of images, colors, animated bodies, voices, props, and sounds is necessary.

Visibility of Recognition

Whenever actors enter the scene in a film, their depictions and performances also rely on patterns of recognition. Certainly, Burton's film too, in his making, to follow James Naremore (1988, 2) in *Acting in the Cinema*, is dependent on "a form of communication whereby meanings are acted out," where, once it is released in cinema, my "experience of watching them involves not only a pleasure in storytelling but also a delight in bodies and expressive movement, an enjoyment of familiar performing skills, and an interest in players as 'real persons.'" Here, the paradigm of the spectacle shifts to the paradigm of recognition as "the interest in players as 'real persons'" remains influential, while the "delight in bodies and expressive movement" becomes central. The film's decision is to offer a basis for orientation from which a "delight in bodies and expressive movement" can develop. The basis for this, the center is the face of Helena Bonham Carter or, to be exact, part of her face, with only some of its features, mostly its eyes, ears, nose, mouth, as Bonham Carter's facial features belong to a disproportionate head that is too big, and are attached to a disproportionate body that is too small. While Hoff describes this partly digital depiction as "a gigantic Elizabeth the first-type head atop a much smaller body," designed in front of a green screen, there are also comparisons in other reviews to a bobble hat atop of a much smaller body. In such shapes, Bonham Carter acts out either familiar or different meanings prominently via her face, while being always ready for the close-up. In her scenes, Bonham Carter's moving, altered face frequently appears at the center of attention whenever she enters and offers both an orientation and a dis-orientation, along with animated body parts that do not belong to herself. Here too, "something sharper than a mask is looming," "something sharper" that Roland Barthes (2004, 721) sees in the face of actress Greta Garbo, "a kind of voluntary and therefore human relation between the curve of the nostrils and the arch of the eyebrows; a rare, individual function relating two regions of the face." However, in Bonham Carter's face, this "human relation" is partly estranged through her digitally enhanced head, supposedly three times its normal size, the prosthetic forehead pushing her hairline back, added by a three-pound heart shaped wig—here, it is interesting to note what the film's visual effects supervisor Ken Ralston (2010) points out when it comes to Bonham Carter's depiction: "We shot her with a 4K camera and really, through brute force, enlarged her head and blended it back onto her body in a way that looks perfectly natural. And once we saw the head that big, we decided creatively to take her waist and do more of an hourglass shape so it balanced that giant head better than leaving her the way she was." While the full body's balancing in an hourglass shape is supposed to keep the spectator in balance during the film experience, the animation's focus on blending an enlarged head with micro-expressions of Bonham Carter's face certainly keeps the possibility of recognition for the spectators in a far from "perfectly natural" realm of experience.

Visibility of Being

Indeed, there is a process of brute force, of exhaustive search in depicting Alice's adversary, the tyrannical, huge-headed Red Queen, prominently through and with the face of Bonham Carter, with an attention to detail carrying affects that initiate experiences of belonging and non-belonging in human and non-human interrelation, at the level of presence, as forces of becoming. From out of the face, its expressions, or to the body parts, up and down, sideways, develops a bodily uneasiness with what is rendered visible on screen, centered on and around Bonham Carter's fragmentized body, following her remaining bodily traces in a slippery realm of film experience, where unresolved affects might prevail, as both the face of Bonham Carter and "her" added animated body parts develop to be un-gripping—causing uncertainties of how to act within the visual, auditory, and tactile field between the screen, Bonham Carter, both her human and digital fragments, Al Hoff, and me, for example. The coexistence with the film actress is challenged, and the bodily link to the actress is loosened. The form which is common to both the actor's and the spectator's visual and tactile perceptions occasionally dissolves, the form that both actor and spectator possess, where, according to Merleau-Ponty (2010, 452), "all happens as if the intuitions and motor performances of the other are founded in a kind of internal encroachment, as if my body and the other's form a system." The form is my body, and what I learn to consider as the other's body is a possibility of movements for me. Thus, as Merleau-Ponty (2010, 453) puts it, "we can say that the actor's art is only a deepening of an art that we all possess." This possibility is partly vanishing from out of the face of Bonham Carter that is attached to a bobble hat and an hourglass shaped body. In addition though, the voice of Bonham Carter certainly develops to be a prominent instrument in depicting her character, it adds to "the visibility of being" an "audibility of being." Bonham Carter's voice impresses us as an incident, whereas both our listening and watching, is always a listening to and a watching at, corresponding to be affected by. We therefore enter the dimension of response, as we respond to what is not ours, what interferes, what is owned by the other.

The film experience then evokes *pathos*. It relates to something that is striking and surprising, that is passionate. With Bonham Carter, our acting subject, her performative body, in expressive modality, and mobilizing affects, we enter an experience that inflicts itself on us—and this infliction appears, while we are responding to it, in between the screen, between her and us, in between the Other and ourselves, neither following a scheme of stimulus and reaction nor a principle of cause and effect, as the effect precedes the cause and points to a gap of impossibility as soon as our own sense-making and sensual possibilities are overcome. But also beyond our acting subject—for example, in following Mia Wasikowska's *Alice* down the rabbit hole through the virtually synthetic Wonderland, meeting creatures, while striving with her for pictures and conversation, where faces do not appear as expected, but enter as something other, in movements, taking place, confronting us, resisting our own subjectivity and for this reason surpassing our own sensual understanding.

NOTES

1. I have questioned these concepts in collaboration with Alice Lagaay and Barbara Gronau at the inaugural conference of the research network Performance Philosophy at the University of Surrey in April 2013: "What Is Performance Philosophy?" The reflections in this section are taken from our co-written and co-organized panel "Ethical Dimensions of the Performative."
2. Within Performance Philosophy, cf. Cull and Lagaay 2014.
3. Cf. Brighenti 2010, 176: *visibility regimes*.

REFERENCES

Barthes, Roland. 2004. "The Face of Garbo." In: *Film Theory and Criticism: Introductory Readings*, edited by Leo Braudy and Marshall Cohen, 589–590. Oxford, New York: Oxford University Press, 6th edition.

Brighenti, Andrea Mubi. 2010. "Artveillance: At the Crossroads of Art and Surveillance." In: *Surveillance & Society*, 7(2), 175–186. Accessed 31 January 2016. http://www.capacitedaffect.net/2010/Brighenti-2010-Artveillance.pdf

Celan, Paul. 1972. "I Can Still See You." Translated by Michael Hamburger. In: *Selected Poems,* 99. Harmondsworth: Penguin.

Celan, Paul. 2003. *Collected Prose*. Translated by Rosmarie Waldrop. New York: Routledge.

Cull, Laura and Alice Lagaay, eds. 2014. *Encounters in Performance Philosophy*. Basingstoke, Hampshire: Palgrave Macmillan.

Debord, Guy. 1995. *The Society of the Spectacle.* Translated by Donald Nicholson-Smith. New York: Zone Books.

Dyer, Richard. 1986. *Heavenly Bodies. Film Stars and Society*. New York: St. Martin's Press.

Frampton, Daniel. 2006. *Filmosophy*. London, New York: Wallflower Press.

Hoff, Al. 2010. "Alice in Wonderland: There are few wonders in Tim Burton's loose adaptation." *Pittsburgh City Paper*, March 11.

Ihde, Don. 1990. *Technology and the Lifeworld. From Garden to Earth.* Bloomington, Indianapolis: Indiana University Press.

Kaplan, David M. 2004. "Recent Philosophy of Technology." In: Readings in the Philosophy of Technology, edited by David M. Kaplan, 89–94. Lanham: Rowman & Littlefield.

Klevjer, Rune. 2006. *What is the Avatar? Fiction and Embodiment in Avatar-Based Singleplayer Computer Games*. Dissertation Dr. polit., University of Bergen, Department of Information Science and Media Studies. Accessed 31 January 2016. https://runeklevjer.wordpress.com

Levinas, Emmanuel. 1996a. "Meaning and Sense." Translated by Alphonso Lingis. In: *Basic Philosophical Writings*, 33–64. Bloomington: Indiana University Press.

Levinas, Emmanuel. 1996b. "Paul Celan: From Being to the Other." Translated by Michael B. Smith. In: *Proper Names*, 40–46. Stanford: Stanford University Press.

Levinas, Emmanuel. 2003. *On Escape*. Translated by Bettina Bergo. Stanford: Stanford University Press.

Merleau-Ponty, Maurice. 1964. "The Film and the New Psychology." In: *Sense and Non-Sense,* 48–59. Evanston: Northwestern University Press.

Merleau-Ponty, Maurice. 2002. *Phenomenology of Perception.* London, New York: Routledge.

Merleau-Ponty, Maurice. 2010. "The Experience of Others." Translated by Talia Welsh. In: *Child Psychology and Pedagogy. The Sorbonne Lectures 1949–1952*, 434–459. Evanston: Northwestern University Press.

Naremore, James. 1988. *Acting in the Cinema*. Berkeley, Los Angeles: University of California Press.

Ralston, Ken. 2010. "Revisiting Alice in Wonderland." In: *VFX World: An Animation World Network Publication*, December 9, Accessed 31 January 2016. http://www.vfxworld.com

Sobchack, Vivian. 1992. *The Address of the Eye: A Phenomenology of the Film Experience*. Princeton, New Jersey: Princeton University Press.

Sternagel, Jörg with Deborah Levitt and Dieter Mersch, eds. 2012. "Etymological Uncoveries, Creative Displays: Acting as *Force* and Performance as *Eloquence* in Moving Image Culture." In: *Acting and Performance in Moving Image Culture. Bodies, Screens, Renderings. With a Foreword by Lesley Stern*, 51–58. Bielefeld, New York: transcript Metabasis, Columbia University Press.

Part II

FACING OTHERS

Chapter 4

Alterity, Machines, and Eros

A New Vision of Communication as an Event

Ciro Marcondes Filho

Communication is possible only with alterity. But not with any alterity. It is necessarily a radical alterity. An alterity that is not just there like an Other, but remains as a stranger in its incapturability, in its eternal fugue. It is also not enough to have an experience of the Other in its impenetrability. It is necessary that the other does not appear by chance, just accidentally, but as expression, or better a saying, that his/her/its appearance in front of me makes me think and changes my stances by creating meaning. A tsunami can change my life, but it is not communication. In communication, the radical interference of the Other is marked by the communicability of this other, as something that comes from a living being, especially the human, whether individual, collective or symbolically, as far as they want to touch and influence me.

So alterity is different from otherness, which can be seen in our relationship with animals, with robots and with nature itself. To them, even in their strangeness before me, this second dimension, the communicability, is not present. Animals and pets do not communicate complex processes, although they may indicate what they feel, what they want, or how they want it or where they want to go. Their verbal understanding it is closed off, they do not capture the broader, abstract context. Machines do not communicate either. They can talk, build contexts and give unpredictable responses but they lack the subtle indeterminacy of the human, especially in Eros, where the one who is closest to me, in an absolute intimate relationship, can flee back into his/her incapturability as a Being that teases me and makes an opening in myself to becoming.

With technology the face of alterity has changed. We do not recognize the robot as having communicability; its otherness is not experienced as an enlightened mystery. Computer-mediated human relations, by contrast, are able to reproduce the radical alterity that Eros would attribute only to direct and in-person interactions.

1. COMMUNICATION AND ALTERITY

Let's talk about phenomenon of communication. But before we get started, we have to agree about the term "communication." Since there is no consensus in this area, the accepted expression of communication is the transmission or transference of messages and the act of making meanings common. In our view this is a mistake, because it assumes that A can pass something to B, when in fact in human processes nothing can be literally transferred. By contrast, we wish to propose a unifying and operational concept. This concept is built from the contributions of Martin Buber and Emmanuel Levinas. Let's examine it.

Communication never happens to a being left alone. It is mandatory to have an Other, a non-me, or a You. Only You can remove me from my isolation. In a brief description about the presence of being-in-the-world, Levinas speaks of three constitutive stages of our existence. In the first, we are "thrown into the world," in Heidegger's sense. We are nothing. We inhabit a so-called force field "exist." We are isolated, an incommunicative monad. Our existence here is sheer loneliness. In a second phase, we are conducted to what he calls the hypostasis: We become existing. We create an identity. We become masters of our existence. We seek to know and learn to own the world. We are matter, consciousness, and subjectivity; but we are not yet able to communicate. For that we must enter a third moment, the confrontation with alterity. It is via alterity that we can come out of ourselves and reach our becoming. We are, therefore, situated in time. We abandon our present and throw ourselves to something which is not us, that will act upon us, and that will hide forever.

In this third phase, I seek not to grasp anything, I do not go in search of enlightenment. The Other, with whom I face myself, will impress me exactly for its incapturability, for its permanent escape. I can never grasp it, and if I do, it will no longer be the Other. At any time I can immerse myself in the Other, for example, in passion when we seek to merge two into one. Love is the opposite of communication. For Levinas this is definitive—the Other is the feminine. Feminine is the same as something else, always other, always inaccessible, always something to become. It is a permanent ambiguity that I will never possess. We can find this feminine in men and women, because it is a philosophical category, and not an attribute of a specific gender. A woman seeking otherness will be faced with the feminine in the man that she seeks; she will want his incapturability, his inexorable escape from herself.

Feminine is the same as becoming. Becoming, not in the meaning used by Deleuze as a common ground where a wasp and an orchid grasp elements of the other without becoming one another. Becoming here is the opening of time, the possibility to get out of our seclusion before the world and enter new possibilities, the unknown, the fascinating, the "infinite" from another reality. The face, says Levinas, is a vehicle to achieve this. Through this vehicle we leave the materiality of the face and reach the human Face, namely, humanity. Looking at a face, we transcend and go out of this presence here and now to reach a metaphysical dimension beyond it.

Levinas also speaks of Eros. In the caresses, this is also possible. Caress is not only feeling the touch of skin or its desire. For Levinas, it "does not know what to search" (Levinas 1946, 82–83). It is a game with something that hides permanently without project or plan. It is not the match of the contact, an undressing that never is naked enough, a closing that does not meet the approach (Levinas 1978, 144). It also has no content. It is a "waiting for the pure becoming" (Levinas 1978, 144) which, according to him, psychoanalysis never realized, since it associates caress always with the pleasure that would be its content.

Consequently, in relation to alterity, the Other is something that goes beyond mere materiality or the carnal. It is a living thing which refers to a spiritual transcendence. It is the invisible visible. A kind of a trace, or "existence that deserted itself "(Levinas 1978, 144). Here we have a brief description of what is our relationship with alterity. More than that, it articulates how this relationship can transform us from being attached to a subject-object relationship with the world, where we seek enlightenment, knowledge, and matter but also domain and possession or the I-It relationship, in Buber's words, to being available to becoming with time, stripped of the need for control, domination and power. In this way, we are open to our own renewal, to that which seems decipherable, an I-Thou relationship. But Levinas goes further, seeking the temporal transcendence of the present toward the mystery of a becoming (Levinas 1946, 89). The figure of a face, a feminine and Eros are several demonstrations of what escapes us all the time, despite its materiality.

Later, Levinas will talk about communication. Communication to him is something that has to do with the dispossession of ourselves. "It is only showing what through which Eros differs from ownership and power that we can admit a communication in Eros" (Levinas 1946, 81). One must leave to be able to communicate, the "I" must be "immolated" (Levinas 1978, 188), trade places with each other, replace each other. The same goes for the Other, without the two relations having the same meaning. Communication is not a two-way street (Levinas 1978, 188). By the time that I put myself "for the other," I am also positioning myself to the other, which is an opening to himself. At this point, Levinas establishes an ethical attribute to communicability. I do not just substitute another, I am also responsible for him. Thus, when looking at the face of the other and seeing in it a human face, I, at the same time, take my part in this story. I become your servant, your deacon.

Finally, Levinas speaks about obsession. When we face alterity something irreducible happens to our consciousness. He says that the subject is affected "without the source of the affection being subjected to re-presentation" (Levinas 1978, 159). Obsession is a kind of relationship that cannot be reduced to consciousness. It is something external, prior to its opening. It crosses the awareness against the flow, signing up on it "as a foreigner, as an imbalances, as a delirium" (Levinas 1978, 159). For this movement, it retrieves the original sense of the word "an-archical." Anarchy, he says, disrupts the being beyond the alternatives conventionally associated with the term. It is a way of how the "I" affects itself and that is a defection of conscience (Levinas 1978, 160).

But there are three drawbacks in Levinas' notion of communication, which we will review in order to propose our concept of communication. First, the notion that communication implies responsibility; second, the fact that femininity as "something that eludes" is associated with female shame (Levinas 1946, 79–81); and finally the association of Eros as full otherness with procreation.

First, when we talk about communication, we assume, necessarily, the existence of the other. It is the one who makes communicability possible. By opening ourselves to another we can leave our solipsism and receive that which is not us. Communication is contacting this stranger that teases us, encourages us, shakes our certainties and, therefore, is able to produce changes in us. This other is often a person, but it can also be a cultural product or something more abstract like a scene, a context, a social movement, or a city. In all these there is, explicitly or not, an intention to communicate. By its presence, by its speech or its image, by their existence, plan to tell me something. It is not like nature, and not at all like robots, as we shall soon see. "To accept the other," however, may have a meaning that is different from Levinas's use. For us it is the relationship with the different and unusual, which is something out of my routine, that appeals to me. He or she appeals to me. They talk to me; they need an answer from me. They do not allow me to stay silent and ignore them. This is what is called responsiveness, which is something very distinct from responsibility, like sympathy with the human Face of the other that does not attribute to me the cause of his condition, which does not mean I do not sympathize. In responsiveness I answer to the other. In responsibility I answer by the other, which I believe is a mistake of Levinas.

Second, Levinas says that femininity as something that eludes is associated with female shame. Here, as well, we notice an unjustified concreteness in explaining the process of ambiguity. Even when dealing with an abstract figure, shame does not cooperate with the reflection. If we want to characterize ambiguity, it would be productive to work with the term—or, with the verb—in its abstract form: the stranger, the unknown, the unusual, the rare, the *unheimlich*. It is evasive precisely because of its inscrutability, its unreachability of its status as the radical other, the never available. Now, female modesty is a cultural form. It is not everywhere nor has it always existed and was as strongly marked by the ascetic rigidity of the Judeo-Christian culture. In current times, when religion declines and people are free to enjoy their bodies, it is not justified to associate the feminine with modesty. Even as the female nude has long transcended the canons of advertising, photographic or cinematographic eroticism, and has become instead the form of its own auto exhibition. The exhibition of the naked female body and its vanity as "compensation for its original sexual inferiority" (Freud 1932, 108) or the "phallic body" (compensation of the absence of the penis [M 2003, 34]) would be today, by contrast, a demonstration of a lack of decency, that even for Freud, was nothing more than a convention (Freud 1932, 108). The contemporaneity, especially after the feminist protests, argues that the freedom of the body, the dis-inhibition, even total hair removal to expose sex, including as a way to deselect it as an object of advertising is an exploration of desire. Sex exhibited as it is, without mysteries and subject to all debauchery that visuality today offers, is not a way to flinch, in which we exempt ourselves from any shameful connotation or even psychic need to cover up whatever it might be.

Third, Levinas works productively with the concept of Eros, but ultimately reduces it to parenting. In other words, he domesticated Eros to the scene of procreation and family. In the three phases of existence exposed by Levinas, where the hypostasis is faced with two outs, death or Eros, the alterity of the female reveals itself as a "victory over death," because in death I have no power. I am not in a position to do anything. I cannot be myself. In alterity, however, as something female, there is a way for me to have a relationship with the other that escapes me but also maintains myself—having a child. Here, the Lithuanian thinker is concerned about the survival of the self. In the face of the alterity of You, how can I stay me, without absorbing myself in You, without getting lost? How can a "I" become another one for you? For Levinas this is only possible in parenting. But we might just as well consider other forms of creation. Levinas says that a child is not a work of mine, but rather something that I produce. It is not something that I have but something that I am (Levinas 1946, 85–86). However, and to disagree with Levinas, the same could also be said about cultural works. I do not have them; they go thru the world. They "occur," or not, living and speaking for themselves.

The notion of a vital elan, which Bergson would have confused, according to Levinas, in the same movement of artistic creation and generation, or what Levinas calls "fertility." This notion of vital elan he says, "does not take into account death but mainly tends to an impersonal pantheism in the sense that does not mark enough contraction (crispation) and isolation of subjectivity, an ineluctable moment of our dialectics" (Levinas 1946, 86–87). In parenting, he says, it is not only the father's renewal in the son or the mix of the father in the child, but it is also the father's externality in relation to the child, a "pluralistic exist." The father comes out of himself as externalized in the child.

But what is of interest in communicability? It is through an Other, my son, that I am opened to becoming, and it is by the action of this very alterity, as Levinas pointed out in terms of "obsession" that something outside crosses the awareness against the fluxes, as an abroad, an imbalance, or a delusion. In my son, as a work, I create another being, who will be, like the others, an elusive otherness, an imbalance and delirium. I produce otherness and I am produced by otherness in different ways. But the circular relationship is inevitable, like the hands of that Escher etching. It changes and turns me to the opening of becoming. It is a way of contracting me, a way to immolate myself, and this is the only possibility for communication. Only when the "I" is immolated, does it contract, says Levinas, in whom the relationship with the other can be communication and transcended (Levinas 1978, 188). And this is what he calls "replacement"—the interior is tearing, the identity is getting lagged. The performing of communication and the relations does not have the same meaning. It is not a two-way street.

But this is the crux of the matter. In communication with alterity, I find myself with the other, I receive him/her/it.; but they always escape me. However and beyond that, they alter me. Something of the other that is left is now going to be part of me. I do not stay the same. This is the I that Levinas wants to protect in paternity as already another, just for being assailed by becoming. There is, then, no other definition for communication.

The figure of Eros is rather innovative in debates about communication. It suggests another perspective in addition to the conventional forms. But, perhaps, it would be interesting to explore Eros as Eros and not reduced to a vehicle of paternity, as Levinas had wished. According to De Saint's review of Levinas, the "intersubjectivity of erotic love leads to an addition of subjectivity to an output of being, which is already transcendence because fertility is inscribed in the depths of erotic desire, it's intrinsically constitutive there—even if modern sexuality insists on separating desire and the pleasure of fertility" (De Saint 2010, 140). But what is proposed here is just the opposite: the dissociation of fertility and erotic desire.

The figure of Eros as the figure of the caress, as view from behind, is an impossible search. The object (the feminine) always escapes me. "In caress, what is there is looked at as if it was not there, as if the skin was the trail of its own retreat, still requesting abatement, as an absence, what however is there, is not there anymore (Levinas, 1978, 144). Also in Eros, the other is there, but she escapes me continuously. The body of the other, like his/her face or his/her eyes, allow me access, but in a totally unknown field that is dark and mysterious. I invade the interior of something indecipherable and, therefore, it fascinates me, because this overcomes my conscience, takes me far beyond it, makes me enter an entirely new territory, opens me to becoming. It is the perfect image of communication as a relation with the stranger. There is no "pleasure of fertility" in this, I do not imagine my continuity in the other, I am simply amazed and delighted by the breaking of my own certainties. And by settling on me it creates a new memory.

2. MACHINES

The confrontation with the radical other strips me of my indifference. It teases me, and it demands an answer from me. It is an Event. Unlike the Event of the philosophers, which is an external happening that runs through me, it demands of me a total recomposition. It messes with my life. In communication, this event is the product of the intentionality of the Other or its expression, which ultimately touches me and makes me feel involved in their drama. But who is this Other? A person. A cultural object. Can it be an animal? Can it be a machine? May it be the natural environment? If we accept that the Other is an agent endowed with intentionality, of the willingness to make me feel, of the determination to get me involved in their drama, or even in their communicative appeal, then we can call it Alterity. That is, the power to communicate. Among humans this is the case of Eros. With Eros we are as close as possible to a person but at the same time more distant. This other in front of me challenges me and will enable me, with its own incapturability, to breaks my solipsism and enter the realm of becoming.

Walter Benjamin would have seen in this exposition about Eros an expression of the aura: the appearance of something distance, however close, what causes it is. But it is the inverse that fits here, possibly the opposite concept, trace, which is the "appearance of a nearness, however far that is what leaves behind; in the trace, we appropriate the thing; in the aura, it takes possession of us" (Kamper 1995, 17–18). In Eros, the other is in front of me, but uncatchable, eternally elusive, but still—or, for that very reason—waking me every time, enlightening me, and/or opening me to infinity. Eros makes it possible for me to reach the extreme possibility through this other person now cleared of social signs—intangible flesh and spirit, giving completely and not ever achieving, opening and closing in its inscrutability of being. It is the distant and different remaining "in assured proximity," as Theodor

Adorno (1966, 191) describes it by way of the concept of the "handsome stranger" of Eichendorff.

Georges Bataille also speaks of Eros, when transcribing the ecstasy of experience in Michel Leiris and his *Mirror of Bullfighting*. Bullfighting, beyond a sport, is a kind of aesthetic expression and, as Leiris puts it, no aesthetic pleasure is possible without some form of rape, trespass, or excess. Sex means, for Leiris, the impossibility of complete communion of two beings, this communion can only take place in death, for example, if the two engaged destroy each other in the climax of the sexual act. As this happens, the "descent" of sexual ecstasy causes disgust, the sickness of a disappointment, a scam, the failure of the non-fusion. But it soon becomes its opposite, because when we pass the fullness of disillusionment, the vacuum that it causes, Leiris says, already leads again to a new delusional aspiration. Consequently, the sexual act would be, in the words of Leiris, the combination of straight and crooked. Sexuality, belonging to the sacred field. It is the "Vertigo summit" taboo, which is at the same time above the ordinary and in the opposite direction of it, prestigious and rejected, a fact which marks in it the same operation of the left and the right. Eros in Leiris is therefore violation and trespass which creates a vacuum, a productive vacuum in himself and for himself.

Next point: a work of art is something endowed with intentionality. It can, therefore, be classified as an Other, even if it is not a human being. At this point, also we also depart from Levinas. An architectural work is likewise an expression of a vision about the world. An aesthetic vision establishing itself as expressivity. Therefore, a city can in itself, by its works, express itself as alterity. It has a face, a social face that speaks.

An animal also has a face, or a visual expression. It informs us of what it wants, for example, food, rest, activity, complicity, etc. It is able to express itself clearly and indisputably, driven by instinctive, biology, or emotional need. There is a kind of speech behind it, but of course there is no intention of changing my conscience or of defending a point of view in order to change my perspective. The animal's eyes can have a pleading look, a fact not unique to humans, and this look can lead me to consider it and to feel for it. By contrast, the people in front of me, who I look at in the face or the eyes, may not be wanting to communicate anything, but they communicate nevertheless. Even if they do not speak, there is still something in them that speaks for itself. And I get emotional. I get involved. I feel, by delegation, their expressiveness. Though animals have direct emotional communication, other humans can refer us to a higher metaphysical ground, to values, morals, or to theoretical and abstract principles not to be found in the animal. For this reason, otherness is an entirely different concept. It is a term that we use when there is no communicative intent.

> I no longer look into her eyes as I hold her in my arms, but I swim through them, head, arms, legs, and I see that behind the eye socket there is an unexplored region, a world of the future, and then there is no logic, just silent germination of whole events, night and day, yesterday and tomorrow. . .The eye, accustomed to the concentration of points in space, now focuses on points in time: the eye sees ahead and in retrospect, according to its own will. The eye is the "I" of myself is gone. This eye without self does not reveal or lights. He travels along the horizon, an incessant traveler, uninformed. (Miller 1946, 177–179)

Robots also have a "face" but a face that does not refer to any metaphysical plane beyond his own presence. If they have communicative intent, I can never see it. Robot eyes lack a pleading look, even if the technology presents emotional stimuli from the machine. Computers, insofar as they function only as a medium or support of human connections, can bind human eyes and even restrict that which has a greater expressiveness.

The fourth type is a kind that is neither alterity nor otherness. Nature has no face but only a way of appearing in its strangeness, its "language," the so-called language of nature.

3. THE CHANGING FACE OF OTHERNESS

Something very curious happens in romantic relationships mediated by the computer, particularly in dating sites. If human opportunities to meet new romantic partners in everyday life boil down to encounters, to contact in the workplace, the university, or other public places; on the Internet the chances multiply in geometric progression and show a multiplicity of hundreds of virtual companions or fleeting friendships. Initial contacts and relationships then unfold and play out not only with an abundance of virtual partners, but also with a another dimension of temporality in the meetings and, because of that, have their own relationship dynamics and developments that arise from them: sympathy, affectivity and sexuality.

On these sites, the other person's knowledge takes place initially through images. The photographs presented, even if outdated, made-up, or not entirely faithful to the suitor's current moment, work as a first appeal or a first exhibition, waiting for the interested parties to visit it. By this photography, I work a kind of roulette, a common term in those environments, in which people successively are "removed" due to this main image being posted on the website. I bypass people and exclude them without mercy and without interest in potential virtues, but only on account of their beauty or facial appearance. This operation is cruel and reveals the disposition of human beings to simply "liquidate" people based on the established standards of beauty and sex appeal.

The next phase is when the person holds in front of them a photograph. During this moment, the traditional ritual of courtship applies, as the applicant can send "winks" to the person who pleased him or even request a personal conversation. In response to these overtures the requested person can, in turn and in the same manner, send an assessment of the person based on summaries of information. The next step, once one has passed through the initial barriers, is the beginning of a dialogue. This is done in writing, whether in a social networking site or through an application like WhatsApp, which many participants pass on without restriction to any candidates in order to develop a more exhilarating conversation. It is also not uncommon for a person to request a video chat in sites that enable this, like Skype or Gmail. One can then have access to other dimensions of communication by hearing the voice and seeing the face by means of a webcam.

Thus we come to the beginning of a personal relationship between two candidates that can remain a mere electronic "friendship" or be transfer to the offline environment, where they will be able to assess their physical contact—to touch the other and experience his smell, his warmth, his presence—and confirm whether the assumptions and affinities created in the virtual environment are consistent with reality. This process no longer interests us and refers to what we discussed above under the concept of Eros. What fits here, in the discussion of the changing face of otherness, is to assess the extent to which Eros also can take place in virtual meetings. The thesis presented here is that yes, in these computer-mediated meetings, before there is any actual physical contact between two people, it can unfold and does so. it is important to point out, with even greater density in the production of sensation, the participation of the body, and the production of meaning and communication.

We assume, in this case, that sexuality is a mental process enabled by the other's body. It is energized by the temporal relationship imposed by the social networking site. This means that, in a short time, there is the possibility of developing a desire and its arising bodily emanations in a compression of time, and therefore more intense than would be needed in a physical meeting. The technology is thus involved in the rhythm of the bodies and provides an acceleration of the processes that ordinarily would need more extended periods of time.

It therefore facilitates the possibility that people not only become available to meet but that these relationships do not suffer continuity intervals, because, in principle, you can be online at any time and in any place—at home, on the street, at work, at leisure. It suppresses the intermediate spaces of emptiness, those broad areas where ideas unfold, mature, and expand. These condensations of time act to speed up the desire and the passion of the movement itself. In doing so, it creates a *sui generis* situation: participants find themselves compelled to enter an activity where speed instigates the emotions, compelling them to participate in a game where the speed of action triggers feelings and irrepressible desires, accelerating the search for direct contact, and emotional and sexual satisfaction. Sex happens to be part of the adventure, as a kind of instant gratification in front of an almost compulsive dynamics.

And it is not necessarily other than sex or sex by another means. Participants, alone in their rooms, can start, by the successive exchange of messages by computer in real time, a true erotic session in which the sequence of actions mimics the actual meeting. One suggests the other to undress, to begin touching herself or touch the partner, and describes in minute details the sequence, so that the other fantasizes when not watched directly by the camera, the sequence of actions that will culminate in mutual ecstasy, encouraged by voice or text from the other—and that also goes for the completion of the act. There is effective enjoyment and effective orgasm. There is the unique experience of being in sexual acts with a new partner, which exposes without inhibiting the reaction of their sexual organs, their movement, their contractions, and their perception of feeling for the other. Indeed penetrating the body and thereby producing the effects of the climax that Leiris spoke of.

There is no reason to doubt that such experiences produce effectively sensations and actual sexual pleasures, especially when one is not talking to sex workers, who stimulate the act but do not necessarily perform it by their own free will. But the emulation produces real effects. The same as one would expect from Eros in communicability: I feel viscerally bound to my partner or she accompanies me throughout the act and encourages me. The other person is with me through images, words, and sounds, but the ambiguity is even more radical in online relationships. It is evident, therefore, that the logic of ambiguity is just a rhetorical, abstract appeal to metaphorize the communication process as an Event. It is a model that materializes the fact that communication is a phenomenon of breaking, of shaking the foundations, or of questioning basic ideas and positions. In short, it is the production of meaning that arises from the experience of a communicational Event. Communication, like Eros, strikes against us, shakes us, takes us out of our tracks, so that we can open ourselves to the different, to that which is not us. It is the confrontation with something beyond our conventional way of thinking, a thought not yet thought, a suggestion that dares to invade our certainties and put them in check. It follows a logic of which we know little and with which we are confronted without being able to register the impact, since the alterity, though disturbing me, always escapes me and will always be strange, unfathomable and mysterious.

If in the meeting of bodies in physical contact I see the excitement on the face of the other, the transformation of his eyes, his face, his whole body in the animal into which such a person transforms itself. Bataille speaks of the woman with extreme elegance who shows evidence of a kind of disease analogous to the anger of dogs: "as if an angry bitch had replaced the personality of this dignified lady" (Bataille 1957, 94). The presence of a being unknown to me, therefore, escapes me and at the same time denounces my inability to decipher her. The same situation that exposes me to what I do not know and that leads me to other scenarios, possibly to the expansion of myself, is present in the electronics that makes it easy to know people capable of exposing themselves sometimes easier than in the actual physical encounter, revealing at the same time the most reserved of us, suggesting a possibly expanded field beyond physical contact.

Violation and transgression, in Leiris's sense, gain a free field when carried out under the protection of electronic devices. If Eros can be used as a paradigm or metaphor of the communicational Event, this means that our exposure to the handsome stranger does not need to face it as it one does in a film theater, or an art exhibition space. Just as the experience of Eros does not necessarily need to be embodied to feel its effects.

REFERENCES

Adorno, T. W. 1966. *Negative Dialektik*. Frankfurt am Main.

Bataille, G. 1980 [1957]. *O erotismo*. 2a. Edição. São Paulo, Moraes.

De Saint Chero, Michaël. 2010. *Entretiens avec Emmanuel Levinas. De la phénoménologie du visage à une philosophie de la rupture*. Paris: Le Livre de Poche.

Freud, S. 1932. *Neue Folge der Vorlesungen zur Einführung in die Psychoanalyse*. Berlin: Fischer.

Kamper, D. 1995. *Unmögliche Gegenwart*. Munich: Fink.

Leiris, Michel. 2001 [1938]. *O espelho da tauromaquia*. Translated by Samuel Titan Jr., 39 São Paulo: Cosac & Naify.

Levinas, E. 1946. *Le Temps et l'Autre*. Quadrige: PUF.

Levinas, E. 1978. *Autrement qu'être ou au-delà de l'essence*. Berlin: Kluwer Academic.

Marcondes Filho, C. 2003. *A produção social da loucura*. São Paulo: Paulus.

Miller, Henri. 1946. *Tropique du Capricorne*. Paris: Ed. du Chêne.

Wiggerhaus, Rolf. 2001. *Die Frankfurter Schule. Geschichte, Theoretische Entwicklung, Politische Bedeutung*. Munich: Deutscher Taschenbuch Verlag.

Chapter 5

Game Over

*About Illusion and Alterity**

Maurício Liesen

This essay is about an illusion of alterity. But this genitive is conceived here in a quite specific manner. The word illusion comes from the Latin *illudo* (to play with/at), which literally means "in-play" (Hoad 1996). Having that in mind, in this text "illusion" refers not only to something which seems to be what it is not, but also to anything that brings something into play, or rather, that points out the structure that creates the play. Consequently, an illusion of alterity can be understood both as a non-performed or fictional experience of alterity and also as an alterity submitted to the logic of play. "Alterity-in-play" may sound like a programmatic study. However, I intend to stress the paradox of this expression. My objective is rather to approach the contradiction of this conceptual relation, because, according to what I am going to argue here, when alterity is defined by means of a play, it should imply the occlusion of alterity's fundamental aspects, namely, irreversability, undeniability, undefinability, uncertainty and, above all, its responsive character.

This theoretical consideration of play and alterity can provide guidance for examining the problematic of communication by way of digital media, which is often described by its interactive—consequently "playing"—features. My argument is that the medial function of play and

*This essay is one of the results of the postdoctoral research project "In the Middle: Obliteration. Advanced Studies in Media Philosophy" (2015/00283-8), supported by São Paulo Research Foundation (FAPESP).

the communicational function of alterity are fundamentally diverse. *Play is a threshold-experience*, whereas *alterity is a limit-experience*. This difference can be exposed by examining digital machines in their relational/communicative forms, since play's mode of being is radicalized in the digital spaces in order to accommodate the game logic, whereas the experience of alterity tends to escape from the mathematical logic of digital machines. To develop the thesis, this text will be divided in three sections. In the first, "A Threshold-Experience," I elaborate the mediality of play and its ontological and ethical entanglements. I also try to summarize the transformations of play through mathematics and digital technologies in order to expose the privileged locus of computer games to think the very idea of digital communication (as interactivity). In the second part, "A Limit-Experience," I discuss the differences between interaction and response and present a concept of communication based on alterity. In the final section, "Defaced Alterity," I highlight the gap between responsibility as effect of alterity and enjoyment as effect of play. My last considerations focus on the problem of using the concept of alterity in the characterization of play experiences based on digital technologies.

1. A THRESHOLD-EXPERIENCE

In a broad sense, play is a free and amusing activity outside the productive and mandatory sphere of work. When play is well structured, with goals and rules, it constitutes a game. So play is the ground of any game. It is what gives the game logic, its mediality, and its mode of being. Moreover, play is a form of experience, which transcends the game space. This mode of relation I call *threshold*—a space "in-between" whose main attribute is to create a secure spaciousness, an interval of the "as-if," where the possibility of experimentation can be guaranteed.

1.1 Mediality of Play

The German philosopher Hans-Georg Gadamer, in his magnum opus *Truth and Method* (originally published in 1960), worked out play's mode of being to highlight its similarities to the being of the work of art: they are closed in an entirely self-contained world (although their self-presentation can be presented to someone else) and they renew themselves all the time (as an eternal back and forth movement). For Gadamer, play and art are experienced as realities that surpass the reality of the player/spectator. To avoid cultural, social and psychological definitions of play centered on players and their actions, Gadamer stressed play's medial function. Citing *Homo Ludens* (a milestone for game studies, published in 1944) by the Dutch anthropologist Johan Huizinga—who conceived play as a free activity which occurs in a sacred space, that is, a space apart from the ordinary life (1980, 14)—Gadamer defines play, above all, as being independent of the player's consciousness (1975, 108 [97/98]).

This primacy of play before the players can be understood as a form of exposing its otherness' structure. And it involves at least two facets: (1) In order for play to *be*, it is always necessary an *other*, in the manner that there will be an answer and then a *set in motion* of the play-structure (Gadamer 1975, 111 [100/101]). (2) Players are rather played than play—they are subjected to the game. "All play is to-be-played": "Unless otherwise indicated, translations are the author's." (Gadamer 1975, 112 [101/102]). The first aspect guarantees the eternal movement of play. It is what produces tension, uncertainty and discovery. The second one attests to the fact that every play is a playing-with and also a dealing with an otherness that cannot be, initially, entirely conceived or conveyed. The fact that "play is the subject of playing" is responsible for an experience of strangeness.

This medial function of play, however, is different from that of art. The French philosopher Roger Caillois elaborates a classification of play and game from Huizinga's work. In his book *Men, Play and Games* (1958) he characterized play as free, separated from the ordinary life, uncertain, and unproductive (as also the work of art), but above all as regulated by rules and structured as make-believe (2001, 9–10). Those two last features provide the foundations of play: rules, orders, and pretendings prescribe and structure the sacred space of play. Artworks, in turn, can reinvent their rules infinitely to the point where no rules are applied—and they will still be an artwork. While play remains within its rules, artworks exceed them. Consequently, the rules and the make-believe disclose the mediality of play. This was already perceived by Huizinga (1980, 22), for whom the distinctiveness of play is "the consciousness, however latent, of 'only pretending.'" Nevertheless, it was the English anthropologist Gregory Bateson, in his 1954 article "A Theory of Play and Fantasy," who developed the ontological and medial entanglements of play and connects it to the notion of lucid dreaming—a threshold between two incompatible but coexistent realities.

1.2 In-lusio

For Bateson, the establishment of all play is based on the meta-communication that "this is play." As he explained, "The statement 'This is play' looks something like this: 'These actions in which we now engage do not denote what those actions for which they stand would denote'" (Bateson 1972, 184). This corresponds to the meaning of a consented illusion. "This is play" operates as something that is slightly (but essentially) different to the symbolic function of "as," which indicates that something stands for another thing. It is rather an allusion (literally "play with"), an evocation of an appearance, but not the thing or the act itself. When kids play that they are fighting, a punch should look like a punch, but obviously not hurt like a real one. It is close to the idea of simulation. Therefore, play's mode of being is distinguished not by a symbolic "as"-structure but rather by an "as if"-structure. As Caillois (2001, 8) pointed out, it "implicates the rules which are behind all games and play activities. It creates fiction, illusions (*in-lusio*)."

The mediality of play *com-poses* not only another reality—like every medium—but also opens a threshold between pairs of incompatible layers, such as reality and fiction, players and spectators, discursivity and performativity. This unique characteristic was exemplarily distinguished by Bateson as the simultaneous equation and discrimination of map and territory. That is why the meta-communication "this is play" constitutes a *frame*, which assures the coexistence of incompatible instances. From the notion of frame, Bateson (1972, 186) highlighted two mains peculiarities of play: "(a) that messages or signals exchanged in play are in a certain sense untrue or not meant; and (b) that that which is denoted by these signals is nonexistent." This is also shared by art, ritual, magic and religion. But it is only "in-play," through its standing meta-communication, that the frame is always experienced as a frame.

Trespassing this threshold implies the end of play. The simultaneity of an inside and outside of play is also the trace of its fragility and its freedom. Play is voluntary: "There is also no doubt that play must be defined as a free and voluntary activity, a source of joy and amusement. A game which one would be forced to play would at once cease being play" (Caillois 2001, 6). Thus, the frame "this is play" not only has ontological but also ethical implications. And the impossibility of its transgression reassures the immanence of its illusion.

1.3 Framed Ethics

Who is the *other* in play? A teammate? An opponent? The play itself? How are actions and manners evaluated? In order to answer those questions, one needs recourse to the structure "this is play," for it is the promise of this meta-layer that ensures not only its existence, but also its ethics. That way the otherness of/in the play is particularly a normative one. It is given by the rules and/or by the make-believe. Consequently, ethical questions of play are attached to its guidelines. Ethical behavior is fair play. It is mutual fairness. If the play order is transgressed, this normative ethic collapses. Good and evil, right and wrong, virtue and vice, justice and crime, they all only exist in the intrinsic disposition of play.

This normative and immanent ethics of play implies that a game can be serious, but its seriousness does not denote responsibility, duty, or even a task. This is also what distinguishes the play from a ritual or a ceremony. That fact was already recognized by Huizinga (1980, 8): "Play is superfluous. The need for it is only urgent to the extent that the enjoyment of it makes it a need. Play can be deferred or suspended at any time. It is never imposed by physical necessity or moral duty." So *superfluousness* of play is, on one side, essential to its freedom and uncertainty, which assures the tension and desire to keep playing. However, the freedom of choice is a *reaction* to the play-structure. The freedom of play demands an answer. But it is free only inside the limits arranged by its rules. In this way, the freedom of choice is at the same time canalized, potentiated, and jeopardized by the order of play. On the other side, the superfluousness of play is also fundamental to the composition of a secure and reversible space. It is what becomes the locus of experimentation *par excellence*. This ability to create worlds inside the world constitutes the pedagogical relevance of play and approximates it to mystic and aesthetic experiences. It enables a temporary cancellation of ordinary life and opens a threshold where one becomes "different and does things differently" (Huizinga 1980, 12).

Naturally, a play's altered time and space are essentially simplified: "The confused and intricate laws of ordinary life are replaced, in this fixed space and for this given time, by precise, arbitrary, unexceptionable rules that must be accepted as such and that govern the correct playing of the game" (Caillois 2001, 7). Furthermore, this other spatiality and temporality are "unproductive," even when the play directly affects the surrounding reality as, for example, in gambling—what is bet is not a product of play. In contrast, what is produced during the play cannot have the same function outside of it. Therefore, excess and waste are attributes of play. As Caillois (2001, 5) explained, a "characteristic of play, in fact, is that it creates no wealth or goods, thus differing from work or art. At the end of the game, all can and must start over again at the same point." This back and forth movement of play is also the possibility of breaking and transforming its rules. Although the other-in-play is already given by its internal logic, the threshold condition of play attests to its persistent openness to an imminent turn. It means also that the otherness of play is not playable; it escapes to it. In the world established by playing there is a framed otherness—an experience of strangeness before a new challenge and then its subsequent overcoming/adaptation.

That is why it is not surprising that mathematical theories of games (Von Neumann and Morgenstern 1944) tried to calculate uncertainty, that is, situations, moves, choices—in short, its decision logic. After the Second World War, the similarity between play and simulation reached the highest level through the calculation capacity of digital machines. Computer games inaugurated a unique era where the framing processes of play and its consequent closeness were radicalized. They can, therefore, be regarded as *the* means of expression of the digital age, for they synthesize its key concepts, namely, interaction, interconnection, control and decidability; and they deepen its ethical and aesthetic potentials. Furthermore, computer gaming can be used as a figure to describe a prevailing way to deal with digital technologies. So an investigation of computer games can be helpful to expose the problematic of alterity in digital media.

1.4 Digital Illusion

In his time, Caillois (2001, 174) criticized the mathematical theories of games for the way they destroy play's reason for being. That is, its contingency. In particular, they attempt to mathematically determine all possible situations and moves in a game. They transform it into information, that is, a decision making within a certain field of possibilities. But what happens if the calculations become the foundation of play itself?

Besides all the main attributes of play—freedom (not obligatory participation), autonomy (own time and space), doubt or tension (that requires active participation of players), and rules—a computer game is also based on cybernetic mathematics. That means that at the heart of any digital game lies a decidability model. According to what the German philosopher Dieter Mersch presented through his deconstruction of cybernetic thought in the book *Ordo ab Chao – Order from Noise* (2014), mathematical formalization is the core of all computer-based devices. So the technological apparatus which enables computer games is founded on cybernetic concepts, namely control, self-regulation, recursiveness, self-reflexiveness, and self-referentiality. This peculiarity was emphasized by many game researchers who have tried to define the particularities of the medium "video game" (Beil 2013; Distelmeyer, Hanke and Mersch 2008; Günzel 2012; Pias 2002; Wolf 2010). Computational mathematics is the reason why the expression "computer game" or "digital game" would be more appropriate than "video game," since they are always subjected to computational logic.

In addition to the intrinsic mathematical logic, from the point of view of the mediality of play, there are, at least, three other differences between digital and non-digital games: the inability to create contingency, the interaction and its consequent immersion effect as the genuine element of digital play to unfold the other spaces of the game, and ignorance about the totality of the rules and the impossibility of their subversion. When it comes to the first, the media theorist Markus Rautzenberg (2015) pointed out that digital play can only produce a "framed uncertainty." This means that computer games articulate a paradox. They simulate uncertainty (normally through randomness), but they cannot create contingency, for they are always inscribed in the game code. As Rautzenberg (2015, 68) explained, computer games "are based in their technical levels on computers that, in the form of nowadays dominant combinatorics founded on von Neumann's architecture and Turing's machines, are not capable of generating real randomness and entropy."

On the other side, the immersive power of computer games consists exactly in creating a self-sufficient world where the player's actions are carried out in spite of the rule system, which creates an illusion of sovereignty over the field of action. Moreover, in order to sustain the feeling of immersion, there are two game modes which are complementary but not coincidental: the performative and the narrative. This categorization was developed by Mersch (2008, 34) in his study of the mediality and logic of digital games. Roughly, this means that when a computer game is played, it is not narrated; and when it is narrated, it is not played. One can assume this literally, because it is quite common that modern computer games have several cut-scenes where the players just watch and which are fundamental to the game's plot.

Nonetheless, these two categories can also be understood as modes of experiencing a computer game, especially since there are games in which the cut-scenes are subtlety integrated into action, like *The Last of Us* (Naughty Dog, 2013). Performance and narrative could be forms of experience in the same way that categories like digital and analog are forms of representation. In his book *Languages of Art: An Approach to a Theory of Symbols* (1968), the American philosopher Nelson Goodman provided a basic distinction between a digital and an analog representational scheme. The former is differentiated or discrete, while the latter is continuous or dense. The German philosopher Simone Mahrenholz (2003) applied those two concepts from Goodman's theory to highlight a difference between discursive and non-discursive thinking. That is, between a sign regarded as an element of a notational scheme or an aesthetic continuum. Regarding computer games, those categories describe coincidental but asynchronous experiences. In more contemplative games, like *The Journey* (Thatgamecompany, 2012) and *Shadow of the Colossus* (Team ICO, 2005), spectator and player's roles are blurred by a different temporarility which is more propitious for turns lacking the need for interaction or reaction. Thus, despite of the mathematical foundation of computer games, there is always the possibility that a game could be also analogically perceived. And this is the case not only for spectators—as Gadamer (1975, 104/105) already pointed out, if a game is presented as a representation for a public, it becomes a spectacle (*Schauspiel*, which is literally, "seeing a game or a play")—but also for the players during the play. Thus, computer games make possible an experience of that double tension between analog and digital representational schemes.

This distancing function is also related to the forms of interaction, mainly through the operation of an *avatar*. It is worth noting that this term, as the concept for the graphical representation of a user within a digital space, emerged as an attempt to make the player feel responsible for the characters in the game *Ultima IV: Quest of the Avatar* (Origin Systems, 1985). The digital persona, like the Hindu origin of the term suggests, should be the incarnation of the player in the virtual world. But between the player and the avatar there is always a distance caused by the device which assures a "secure" experience. The avatar is never the player himself/herself and yet stands for him/her, but only within the play world, as a kind of mask. Also, the avatar produces neither a specular nor identity relation. It is already something other than player. Even in first-person games, there is always an asynchronous and dys-topic perspective where the avatar is perceived/observed at a distance. There is, then, a persistent distance taking through the technical apparatus and, especially through the experience of that phenomenon as play, a threshold-experience. The avatar is not lived, but rather controlled, tested, in short, played. There may be empathy, identification, but hardly an experience of being someone else.

Moreover, as a calculated and given world, the player does not need to know the rules to be able to act in it, as is the case in non-digital play. For that reason, some scholars argue for the impossibility in computer games of a *magic circle* (Liebe 2008)—an Other world inside the world, but with fragile rules that can be interrupted at any time—or even of the meta-communication "this is play" (Neitzel 2008). Furthermore, the players cannot break or change the rules during the play, for they were already preprogrammed. For this reason, the mathematization of play, its continual capacity for repetition and the consequent unimportance of being aware of the rules implicate at least two ethical consequences. First, contrary to the computer games theorist Jesper Juul (2013), who argues that computer games are a proper medium to experience failure, the reversibility and recoverability of digital play preclude the sense of irreparability, loss, and trauma. When someone loses a game constantly, there is not inexorable failure, but frustration. The irreparable is outside of play. If it occurs, it is always extrinsic to it. What is irreparable lays on the materiality of play; it is what escapes its logic.

Secondly, the gradual irrelevance of the meta-communication "this is play" transforms digital playing into a relational regime, which can be applied to other forms of mediated social relation, such as social media, education platforms, and dating applications. If, on one side, this regime is effective to achieve goals, that is, to learn and improve skills, there is, on the other side, the incapacity to distinguish "playing" from "not playing"—and according to Bateson (1972, 197), this is also a symptom of schizophrenia—producing irresponsibility, or rather, non-responsiveness. Play demands reactions so that it continues and goes on. But interaction is not the same thing as response. The reversibility of the actions and the experimental character prevent the uniqueness of the encounter with the radical Other. The digital logic takes the threshold-experience of play to its extreme: being neither inside nor outside and reassuring always the possibility of return (without any "scratch") to inside, to the *Same*. As long as ethical responsibility before a radical other is the consequence of an experience of alterity, it would be impossible to have this occur in a computer game. In other words, when communication assumes a play-structure, it can denote communication but it does not present what communication would present, that is, alterity. The consequence of this difference is the subject of the next section.

2. A LIMIT-EXPERIENCE

Different from the regulated and experimental spaces of play, alterity implies a form of relation beyond any rules, control, or repetition. Above all, alterity is an experience of the irreparable. It exposes a relation to an "outside of everything," whether it be called exteriority, transcendence, infinity, or the Other. It does not mean a relation to another structured world opened in everyday life—as if by playing—but rather it is the limit of the experience itself before an Other who escapes understanding and withdraws from the power of the subject of experience. *Ergo* alterity is a relation that culminates in its own impossibility. To use an expression developed by the French-Lithuanian philosopher Emmanuel Levinas, who definitely inscribed this word in the literature of Western philosophy, alterity expresses a "relation without relation" (1979, 80). It means that this relation in fact occurs, but cannot be apprehended; it indicates the limit of all human experience.

Here, I would like to draw a distinction between otherness and alterity: the difference depends on the latter's radicality. While "otherness" can be used to describe an Other who can be assimilated through the experience—like the expanding circles of play—"alterity" is irreducible to interiority or any hermeneutic process of assimilation. Otherness can describe the other during the "this is play." But alterity is what grounds communication. It is an ethical experience, not because it forms a code of rules or prescriptions for how to act, but rather, because it is an appeal, a calling into question that, in the presence of the radical Other, exposes the insufficiency of the subject and opens the possibility to communicate.

2.1 From Interaction to Response

Interaction is the constitutive element for human-computer relations. It defines the logic of action-reaction or input-output through which an intentional subject "communicates" with the machine. Interaction is also constitutive for the play. It sets in motion the dynamic of the game. Therefore, interaction presupposes intentionality, for it describes a process in which a subject turns something into its object by acting intentionally. So in the constituted world of the game, in which the action of the player is presupposed, nothing happens without an intentional act. Even before the advent of cybernetic theory, in the social-psychology belonging to the early twentieth century, the concept of communication corresponded to the notion of interaction. In this way, communication was understood as the social interaction of an individual with another through technical means of transmission and information. This "communicational turn" (in American sociology) was consolidated by the Chicago School (Cooley 1909; Mead 1934; Sapir 1935). Before the commercial usage of computer games, interaction and communication were already regarded as synonyms and the play-structure of rules, players, and actions was applied to describe the experience of communication.

But such equivalences between the concepts of communication and interaction had already been questioned. More than thirty years ago the French philosopher Jean-François Lyotard, in his essay "Something like: Communication. . .Without Communication" (1991, 108 et seq.), criticized the use of the concept of communication to characterize interactive artworks. According to him, intentionality and control have nothing to do with communication, for communication demands a passibility that constitutes a communal experience as a non-conceptual or immediate one, that is, founded on alterity. The starting point for his consideration is the analysis of the sublime according to Immanuel Kant, more precisely the fortieth paragraph of the *Critique of Judgment*, entitled "From the taste as a kind of sensus communis" (Kant 1991, 213–218), where the philosopher wrote that taste is an *a priori* faculty of judgment of what strikes the senses and that it derives from a universally communicable representation without the mediation of concepts. A non-conceptual or immediate communication is incompatible with the notion of communication as selection, transmission, and control of information. It is even more incompatible with the concept of a safe experience or experiment, and thus also incompatible with the concept of play. Reflecting on artistic productions that extol the primacy of interaction, Lyotard questions the place of passibility in the communicational experience of artworks. This passibility prior to any control and intentionality is what characterizes the opening for the communication experience. That opening is alterity—a radical ethical face-to-face encounter, which is completely inapprehensible, unpredictable, and cannot be operated, controlled or switched-off. I can only respond. Different from the logic of play, alterity assumes a logic of responsiveness.

Responsiveness is a constituent of any experience of strangeness. Strangeness is what is out of any category, system, or order. It is the extra-ordinary. Therefore, the process of accommodating an experience of strangeness to a regime (via naming, classifying, locating, explaining, etc.) is normalization. However, according to the German philosopher Bernhard Waldenfels (2008)—who in contemporary philosophy, in a programmatic and deeply way, reflects on the experience of strangeness—this normalization process never actually reaches the strangeness itself, only its effects. So if one considers play under the viewpoint of dealing with strangeness, the Other is incorporated, normalized, reduced to the Same through the expanding circles of comprehension. It is the hermeneutics of play. This process of returning to the Same is what Levinas (1979, 110 et seq.) called enjoyment (*jouissance*). On one side, this expression can be understood literally: normalization is pleasurable, it is like a return to home. On the other side, enjoyment means closeness to the other, to a wandering in the world. Let us take a close look at Levinas's own words:

> In enjoyment I am absolutely for myself. Egoist without reference to the Other, I am alone without solitude, innocently egoist and alone. Not against the Others, no "as for me . . ."—but entirely deaf to the Other, outside of all communication and refusal to communicate—without ears, like a hungry stomach. (1979, 134)

In other words, to affirm the radical nature of the other—that is, alterity—is to say that it cannot be accommodated to the self or be suspended by generalizations. According Waldenfels (2006, 46 et seq.), the experience of strangeness has basically two dimensions: the rapture and the response. The rapture describes how the experience of a radical other occurs: our attention, despite any intentionality, is so to speak assaulted. It marks the uncontrollability of the time and the space of the experience. Something just takes us, it happens to us. In the rapture, the subject is no longer subject, but rather subjected (to). The rapture is a precedence that requires the priority of a response. The rapture and the response are not two successive events—not even two events. They describe a single time-displacement experience. This experience is above all ethical, because it features not only a reception, but an answer.

This answer has its own logic, which is different from the logic of understanding (in the sense of hermeneutics), from the logic of communicative action (discourse ethics), and from the logic of the intentional act (phenomenology). Waldenfels identifies four elements which characterize the logic of answering or responsiveness: (1) Uniqueness. That is, the singularity of events, which allow other ways of seeing, thinking and acting. (2) Inevitability, because you cannot not answer; the not answering is already an answer. (3) Displaced temporality. "The original delay of response overcomes the primacy of an original present. . . . The answer happens here and now, however it begins elsewhere. . . Responding means do without a first—and consequently also without a last—word" (Waldenfels 2006, 65). (4) Asymmetry. The experience of strangeness, for Waldenfels as well as for Levinas, does not establish any dialogue, but rather a rapture, which is considered to be the moment of radical passivity. Thus the logic of responsiveness constitutes the mediality of communication. Consequently, responsiveness precedes intentionality and founds an ethical relation that always demand a responsible answer.

2.2 Being Outside

According to what I have presented so far, alterity has nothing to do with certainty, coherence, and agreement, but concerns disruption, proximity, and responsibility (considered, above all, as an inevitable response). On the fringes of the experience of communication, the encounter with the radical Other is made possible. That is the reason why the limits of communication are at the beginning of the ethical relation. No wonder, then, that communication became a philosophical problem only when existence—an ethical existence—emerged as a question. First, by the Danish philosopher Søren Kierkegaard and later by the German Philosopher Karl Jaspers, who was the first to consider communication to be a proper philosophical category. According to Kierkegaard (1980; 1997), communication has, fundamentally, an ethical function, for the real communication can only be indirectly communicated. In the manner of the Socratic maieutic, communication is an ethical gesture to help someone else find his truth. For Karl Jaspers (1973), communication assumes an ontological and existential function, for Being can only be a Being in communication. After him, Martin Heidegger would crystallized this constitution of Being from alterity in his concept of Being-with or *Mitsein*.

But it was in postwar French philosophy where the concept of communication achieved its radicality. That was caused by an intense intellectual dialogue, especially between Georges Bataille (1962; 2014), Maurice Blanchot (1993; 1988), and Jean-Luc Nancy (1991; 2000; 2002), whose books are direct answers to the previous work by one another (for instance, one can see this clearly in the discussion about the re-signification of the concept of community). As for the concept of communication, Bataille (2014, 96 et seq.) perceived it as an "inner experience." The adjective "inner" does not mean a subjective experience, but an intimate and profound one. Such an experience occurs at the limit of the subject. It is an experience that questions the primacy of the subject and its intentionality. Bataille also called it "ecstasy" (from *ekstasis*, "to stand outside oneself"). For this reason, death stands as the non-experience *par excellence.* My death is something I cannot experience. It is then the death of the other, that reveals to me the impossibility of knowing anything about it. In other words, exposure to death exposes the impossibility of communion. It is an exteriority that produces a common non-belonging.

To stress this experience of exposure (of the subject's limits), Blanchot (1993, 202 et seq.) used the expression "limit-experience" in his discussions of the work of Bataille. Also a reader of Levinas, Blanchot emphasized the role of alterity—or death as something beyond conveyance—as the foundation of this experience. More recently, Nancy—whose concept of communication oscillates throughout his entire career between different words, such as *partage*, *Mitsein*, and community—defines communication by its transcendentality, but in a very specific sense: "The being-communicating (and not the subject-representing), or if one wants to risk saying it, communication as the predicament of being, as "transcendental," is above all being-outside-itself" (1991, 24). In this sense, communication is understood as violence against the current meaning of this word, be it as subjectivity, intersubjectivity, interaction or transmission.

Based on this rough outline of the philosophical approaches to communication, I want to emphasize that the communicational thinking of ethics, aesthetics, and alterity—that is, beneath or beyond informational techniques and technologies—has already a long career outside of communication sciences. Furthermore, my intention was to show that an experience of communication—no matter how it is called—describes a movement of *ek-stasis*, a desire without possibility of fulfillment. In other words, it is a relation to an exteriority that implicates a responsibility for someone who surpasses me, who exposes my insufficiency and my finitude before his/her infinity. Above all, it is a radical experience of humanity. For this reason, communication is to be exposed. It is a concrete situation before an Other, who demands response and respect. It is a promise to overcome the solitude, but not the isolation, of time-space distance, by being obligated to respond to this always preceding demand. In this sense, responsiveness is the very groundless ground of communication. It is the foundation of a communicational motion which does not refer to any origin. As Waldenfels (2006, 67) wrote, "We invent what we respond to, but not to what we respond to and what lends weight to our words and actions." Therefore response, respect, and responsibility are not only in the same semantic field; they belong to the same phenomenon.

3. DEFACED ALTERITY

Understood as relational forms, play and alterity are not opposites but different—or rather, separated. The communicational experience of alterity obliges us to face an ultimate responsibility, whereas the medial experience of play brings us enjoyment through the expanding circles of learning and assimilation. Alterity defines a very concrete experience which founds the social and can be understood under the re-elaboration of the concept of community. This word, in particular, has undergone a theoretical turn since the second half of the twentieth century. After the fascist concept of community grounded on blood and land, and after the communist idea of a being-together based on work, there was an attempt to redefine the concept of community, especially by French postwar thinkers. The works of Bataille (1962; 2014), Blanchot (1988), Nancy (1991; 2000) and the Italian philosopher Roberto Esposito (2010) provide another concept of community not defined by a possession or by a common feature, but rather as what is beyond or above any intersubjective relation. A community founded on alterity asserts that an independent individual is unattainable, for the subject can only be on account of an Other. Because of it, community is revealed through our condition as mortal and insufficient beings. The Other's death makes possible both the impossibility of communion and the revelation of mortal beings' community—a community without any possibility of belonging.

To be clear, let us take a quick look at the work *Communitas: the Origin and Destiny of Community* by Esposito, in which he inquires about the origin of the word "community" in the Latin *communitas*. This term is basically made of two words: *cum* (with) and *munus* (a burden, a debt, a task). Among the variety of meanings of the term *munus*, Esposito emphasizes the sense of "to owe something"—as a duty, as something that is imposed, as an onus, to answer. As the philosopher explains, the *munus* "indicates only the gift that one gives, not what one receives. . .The *munus* is the obligation that is contracted with respect to the other and that invites a suitable release from the obligation" (Esposito 2010, 5). Therefore, such forced donation implies a loss and a grant, not a sharing of something owned. You give something that cannot keep to yourself, even in the mutuality of giving. For this reason, *communitas* characterizes the encounter of people who are close not by "property" but by a duty or a debt. The communicational motion is promoted by an insufficiency, by an original negativity. Those who are, by contrast, free from this demand and obligation are immune, they are called in-dividuals. Debt and giving, therefore, correspond to the responsive feature of alterity. In this way, communication implies not only responsiveness, but responsibility and respect.

However, when it comes to communication by digital media, there is a persistent risk that communication turns into an information exchange, that is, when the responsive structure gives in to the computational logic of play. By losing the meta-communication "this is play," digital play is not experienced as a frame anymore. That fact opens the playful potential of digital media—no wonder that in the last decades the digital play-structure has exceeded the boundaries of computer games to the point that "gamification" has become an important concept in social media, conflict resolution, training program, educational processes, and businesses (Beil 2013; Fuchs et al. 2014; Kapp 2012). Nevertheless, when communication is "played," it loses its responsive features. By the "illusion" of alterity there is no radical Other, but rather players or objects to be played. The framed encounter with others means that the subject is always in control. For instance, the secure spatiality of regulated and reversible play can support discourse that would not be spoken by face-to-face communication or in front of an audience (e.g., hate speech and defamation). Furthermore, the reward structure and the circularity of the game logic (e.g., *like* and *recommend* buttons) provides a form of communication devoid of responsibility. It is not surprising that when an offensive or even criminal post is published in a social network and attracts media attention, the first thing the post's author does is attempting to delete it, as if the structure of the game's reversibility could be applied to social communication. However, there is no "reset button" or "undo" option in communication, only reparation, after taking account of responsibility.

All the clicks, drags, and swipes—all of which are very common in dating apps—resemble simple decision structures in digital games. A person's profile is not representing the person; it constitutes an avatar that can be chosen or swept away. This has nothing to do with the encounter of the face—in the sense of Levinas' *visage*, as epiphany, expression and source of meaning. Rather it is the perception of an object, of a "pic," which I can obliterate or eliminate without any responsibility before it. I do not intend to perform here a pessimistic diagnosis of the use of these technologies. Nor do I intend to say that these tools could not be used differently. I just want to point out that the logic of play is different from the alterity responsiveness and that the lack of the meta-communication "this is a game" is the cause for many forget that the logic behind social media is not communication but digital (mathematical) play. In this way, alterity is the key to understanding what is unique in face-to-face communication.

As discussed in the first section, enjoyment, which is the main effect of play, is self-sufficient. Thus players do not want to be disturbed. In turn, the communicational experience is exposure to a risk and the duty to answer without question. It is an original answer to the presence of the Other. Thus, the irreparable is always already outside of play.

Radical alterity brings the question of sharing instead of participation, of communication as revelation of existence and of a community that demands a response that cannot be given or undone. It is an ethical experience, occurring during the encounter between human beings, which founds sociability and opens the possibility not to be other but to be exposed to others, to be outside of oneself. No wonder, then, that Levinas hesitated to assign alterity beyond the face-to-face encounter, as he assumed in his essay "Is Ontology Fundamental?" (1996). The other paralyzes my own power to power. Alterity is an ethical experience of responsibility before an Other who surpasses me infinitely and who opens me up to the externality of my existence. When it happens, the game is already over.

REFERENCES

Bataille, Georges. 1962. *Erotism: Death and Sensuality*. Translated by Mary Dalwood. San Francisco: City Lights Publishers.

Bataille, Georges. 2014. *Inner Experience:* Translated by Stuart Kendall. New York: State University of New York Press.

Bateson, Gregory. 1972. "A Theory of Game and Fantasy." In *Steps to an Ecology of Mind*, 183–199. Northvale-NJ/London: Jason Aronson.

Beil, Benjamin. 2013. *Game Studies–Eine Einführung*. Berlin: Lit.

Blanchot, Maurice. 1988. *The Unavowable Community*. Translated by Pierre Joris. New York: Station Hill Press.

Blanchot, Maurice. 1993. *The Infinite Conversation*. Translated by Susan Hanson. Minneapolis/London: University of Minnesota Press.

Caillois, Roger. 2001. *Man, Play, and Games*. Translated by Meyer Barash. Chicago: University of Illinois Press.

Cooley, Charles Horton. 1909. *Social Organization : A Study of the Larger Mind*. New York: Charles Scribner's Sons.

Distelmeyer, Jan, Christine Hanke and Dieter Mersch, eds. 2008. *Game over!? Perspektiven Des Computerspiels*. Bielefeld: transcript.

Esposito, Roberto. 2010. *Communitas: The Origin and Destiny of Community*. Translated by Timothy Campbell. California: Stanford University Press.

Fuchs, Mathias, Sonia Fizek, Paolo Ruffino and Niklas Schrape, eds. 2014. *Rethinking Gamification*. Lüneburg: Meson Press.

Gadamer, Hans-Georg. 1975. *Wahrheit Und Methode: Grundzuge Der Philosophischen Hermeneutik*. Tübingen: Mohr Siebeck.

Goodman, Nelson. 1968. *Languages of Art: An Approach to a Theory of Symbols*. Indianapolis/New York/Kansas City: Hackett Publishing.

Günzel, Stephan. 2012. "The Mediality of Computer Games." In *Computer Games and New Media Cultures*, edited by Johannes Fromme and Alexander Unger, 31–46. Dordrecht: Springer Netherlands.

Hoad, T. F. 1996. "Illusion." *The Concise Oxford Dictionary of English Etymology*. Oxford University Press.

Huizinga, Johan. 1980. *Homo Ludens: A Study of the Play-Element in Culture*. London, Boston and Henley: Routledge & Kegan Paul.

Jaspers, Karl. 1973. *Philosophie II. Existenzerhellung*. Berlin/Heidelberg/New York: Springer.

Juul, Jesper. 2013. *The Art of Failure: An Essay on the Pain of Playing Video Games*. Cambridge: The MIT Press.

Kant, Immanuel. 1991. *Kritik Der Urteilskraft*. Edited by Gehard Lehmann. Stuttgart: Reclam.

Kapp, Karl M. 2012. *The Gamification of Learning and Instruction: Game-Based Methods and Strategies for Training and Education*. Hoboken: John Wiley & Sons.

Kierkegaard, Søren. 1980. *Entweder-Oder: Ein Lebensfragment*. München: Deutscher Taschenbuch Verlag.

Kierkegaard, Søren. 1997. *Die Dialektik Der Ethischen Und Der Ethisch-Religiösen Mitteilung*. Bodenheim: Philo Verlagsgesellschaft.

Levinas, Emmanuel. 1979. *Totality and Intinity: An Essay on Exteriority*. Translated by Alphonso Lingis. The Hage/Boston/London: Martinus Nijhoff Publishers.

Levinas, Emmanuel. 1996. "Is Ontology Fundamental?" *Basic Philosophical Writings,* edited by Adriaan T. Peperzak, Simon Critchley and Robert Bernasconi, 1–10. Indianapolis: Indiana University Press.

Liebe, Michael. 2008. "There Is No Magic Circle: On the Difference between Computer Games and Traditional Games." In *Philosophy of Computer Games*, edited by Michael Liebe, Dieter Mersch and Stephan Günzel, 324–341. Potsdam: Potsdam University Press.

Lyotard, Jean-François. 1991. *The Inhuman: Reflections on Time*. Translated by Geoffrey Bennington and Rachel Bowlby. Cambridge: Polity Press.

Mahrenholz, Simone. 2003. "Analogisches Denken. Aspekte Nicht-Diskursiver Rationalität." In *Die Medien Der Künste: Beiträge Zur Theorie Des Darstellens*, edited by Dieter Mersch, 75–92. München-Paderborn: Wilhelm Fink.

Mead, George Herbert. 1934. *Mind, Self and Society from the Standpoint of a Social Behaviorist*. Edited by Charles W. Morris. Chicago: University of Chicago.

Mersch, Dieter. 2008. "Logik Und Medialität Des Computerspiels. Eine Medientheoretische Analyse." In *Game over? Zur Medialität Des Computerspiels*, edited by Jan Distelmeyer, Christine Hanke and Dieter Mersch, 9–41. Bielefeld: transcript.

Mersch, Dieter. 2014. *Ordo Ab Chao–Order from Noise*. Berlin/Zürich: Diaphanes.

Nancy, Jean-Luc. 1991. *The Inoperative Community*. Minneapolis: University of Minnesota Press.

Nancy, Jean-Luc. 2000. *Being Singular Plural*. Translated by Robert D. Richardson and Anne O'Byrne. Meridian (Stanford University Press). California: Stanford University Press.

Nancy, Jean-Luc. 2002. *Hegel: The Restlessness of the Negative*. Translated by Jason Smith and Steven Miller. Minneapolis: University of Minnesota Press.

Neitzel, Britta. 2008. "Metacommunicative Circles." In *Philosophy of Computer Games*, edited by Michael Liebe, Dieter Mersch and Stephan Günzel, 278–295. Potsdam: Potsdam University Press.

Pias, Claus. 2002. *Computer Spiel Welten*. Munich: sequenzia.

Rautzenberg, Markus. 2015. "Gerahmte Ungewissheit. Medialiät Und Kontingenz Im Digitlen Zeitalter." Edited by Christoph Wulf and Vera Zabotkina. *Paragrana* 24 (2), 57–73. Berlin: De Gruyter.

Sapir, Edward. 1935. "Communication." *Encyclopaedia of the Social Sciences*. New York: Macmillan Publishers.

Von Neumann, John and Oskar Morgenstern. 1944. *Theory of Games and Economic Behavior*. Princeton: Princeton University Press.

Waldenfels, Bernhard. 2006. *Grundmotive Einer Phänomenologie Des Fremden*. Frankfurt am Main: Suhrkamp.

Waldenfels, Bernhard. 2008. *Grenzen Der Normalisierung. Studien Zur Phänomenologie Des Fremden, Bd. II*. Frankfurt am Main: Suhrkamp.

Wolf, Mark, ed. 2010. *The Medium of the Video Game*. Austin: University of Texas Press.

Chapter 6

Facebook and Rolezinhos

Alterity, Communication, and Visibility

Alexsandro Galeno

For Tati

This chapter will show that connectivity, faciality, alterity, and spatiality need to be understood as ontological and epistemological categories of the phenomenon of communication. Is it possible, in the contemporary digital era, to think of communication without a connection between the subject of communication and the media of subjectivity? What visibilities are produced in this relationship? Consequently, the important philosophical question is not only "what are we?" but "where are we?" and "how are we, when interacting with technical devices?" When interacting online, we are inexorably connected to the world of other things and human beings. If it is true that with the advent of virtual communication, faces often become invisible or metamorphose into the carefully constructed and controlled profiles of social media, then we might ask what becomes of the concepts of the public sphere and of social and political action? Very distant from the Luddites, who sabotaged technocratic transformations by breaking the machines, today we are witnessing other forms of revolt, like that of the Rolezinho, a social movement of young people from São Paulo's periphery, which mainly occurred in 2013. How do we classify this form of opposition? Especially because they have their meetings and demonstrations at the center of hegemonic control: in the shopping centers, the malls, and the Crystal Palaces of consumption. This chapter considers the case of the Rolezinhos, who use the technology of the Internet, Facebook, and mobile devices as prostheses of sociability to form oppositional communities engaged in what appears to be the will to nothingness.

1. INTERNET: A NEW HUMAN COMEDY

A New Theory of Communication (NTC) can be derived from literature to problematize ontological and epistemological questions. In the works of either Balzac or Proust, we experience events that provide images and communicate an aesthetic sense that expresses the condition of the subject in the world. It is in this direction that NTC, elaborated by Ciro Marcondes Filho, proceeds. It is an epistemological proposal to rethink communication in such a way as to emphasize literary, artistic, and philosophical aspects:

> For Proust, according to Gilles Deleuze's reading, signs faithfully represent the scope of communicability; they 'violate' the thought and this is what we need to investigate in communicative practices. Adding to this, we say that they make us think, force us to think, nothing more than that. This is communication. In Proust's opinion, only through art (and in our case, also for communication research) we can get out of ourselves, get to know what the other sees from his/her universe, which is not ours. We can, in a word, investigate human incommunicability both in person and in front of systems and virtual communication technologies. (Marcondes Filho 2012, 77–78)

Ideas, as defined Balzac in his autobiographical novel *Louis Lambert* (1993), are like Nature's blooms. Would Balzac not have, with his *La Comédie Humaine (The Human Comedy)*, anticipated one kind of hypertext communication, considering that he created the narrative method of the reappearance of characters in different novels? Novels that connect and link themselves infinitely. Can the rhizomatic universe of *La Comédie* contribute to a new epistemology through its recursive and reconnecting model between distinct kinds of knowledge? Balzac, it is believed, retrieves the etymological sense of the word *complexus*, which weaves together, and relates it to the ethical appeal of the words *complexere* (embrace), *communicatio* (communication), and *communicare* (communicate, be in communion). Epistemological clues in this direction have been presented by Franco Moretti's research, which has fabricated networks of characters, plots, and actions:

> For those who work with novels and theatrical plays, style is only part of the picture. What's in the plot? Can we quantify it? [. . .] It is a theory that allows us to analyze the connections between large groups of objects (any object: banks, neurons, film actors, investigations, friends . . .) to which we call 'knots' or 'vertices' joined to one another by the so-called 'edges' or 'arcs'. [. . .] A network consists of vertices and edges; of plot, of characters and of actions. (Moretti 2011, 71–72)

Balzac, undoubtedly, contributes to a NTC and evokes innovative studies in the field of Network Theory through the connection or interactive communication present in *La Comédie Humaine (The Human Comedy)*. Such evidence is proved by the amount of characters—2472 according to Marceau (1986). From these, 573 reappear in several novels or narratives. Remarkable are the following: Rastignac, Bianchon, Nucingen, and Henri de Marsay. *The Human Comedy* as a kind of interactive social network? In this particular way, would Balzac not have anticipated in some way the efforts of Mark Zuckerberg and the invention of Facebook as a network that articulates characters, events, and narratives? Through the stylistic feature of the return of characters, Balzac built a network of unity and relinked narratives that comprise his masterpiece. He instituted a unique form of connection or communication that make up and support the book. He populated a virtual world, creating crowds in his social and historical narratives.

2. THE ENIGMA OF SPACES, ALTERITIES, AND ONLINE SCENIC MASSES

In the post-media-book era, another comedy has just begun. This is a kind of communication made from the connections between people and new devices. We believe that the new Comedy is made from the idea of connection, which is essential to interactive communication. We recall, for instance, that the word connection comes from Latin *connectare* (tie together or tie one another). The addition of the prefix *com-*(together) to *nectere* results in "connect" and "tie," and from this root came the word "nexus." Thus, to connect is to communicate (*comunicare* or "to be in communion").

This new comedy made possible through the interconnections of new media devices and channels expands the place of writing and of the subject. As an example, you only need to cite the proliferation of posts on Facebook and tweets on Twitter. In a simple Twitter message, or "tweet," we can be linked to images, sounds, and other messages. By streaming and the logic of interactivity, we witness the supremacy of a single subject that enunciates, but this time, he/she is accompanied by a crowd of other subjects that are his/her "followers." Therefore, it is impossible to have an experience of alterity where the Other is disconnected, separate, or isolated in his/her online agora. Even though many feel each one is located at the height of his/her own status as a kind of sovereign narcissus, it is impossible not to "tie" or not to "connect" to the Other. This does not mean that one will have a commitment, *a priori*, of being in communion with ideas and/or similar belongings. When connecting, we are inexorably joined to the world of things and men. We are nothing without these things and men! We are connected, even when, as is sometimes the case, we start from a condition of the alien or stranger, who is strange, but not separated. This is an alterity that expresses, in a radical way, the tie between them, a kind of a radical alterity in which "I is an Other" (Rimbaud 1997) and therefore tied, linked, and (inter)connected.

Freud, in his psychoanalytical theory, considered the enigmatic character of man's relationship with space. Such an enigma puts to the subject the continuing need to ask where s/he is to be found. Although we should not disregard the Freudian position, Sloterdijk warns that such questions are not restricted to psychology, but involve a new theory of space, comprised of alterities exerted by man and his spatial configurations. Looking at the contemporary era, then, we can ask what topological enigmas and alterities are we to decipher? One of them is most certainly the different forms of alterity on the Internet. We now perceive the appearance and production of images from what we might call the Online Scenic Masses.

Online Scenic Masses redefine some of the classic categories of the philosophy of space and communication as well as the place of the political and the communicational event. One of the redefinitions concerns the idea of alterity produced among subjects that are manifested. They are a kind of post-gathering social mass. They are media masses that make use of technical devices in the Web 2.0 era, and they are organized from platforms such as Facebook and Twitter. Moreover, they are in-connected masses, because they operate by horizontal communication and by the flow of the event or as "Things crossed by a *pneuma*" (Marcondes Filho 2013, 42). By this, he means, "the breath of life" and the strength and energy that constitute the Event, which is the fundamental basis for his theory of communication. "There are no beings," Marcondes Filho (2013, 41) writes, "only events. This suggestive statement, which come from the interpretation of the thought of Heraclitus, is the foundation of our theory of communication, of both interpersonal and of the masses through the Internet." Without leaders, *a priori*, the masses of the Internet move like mediatic waves forming swarms in a hypersphere of interconnected components. In this sense, if they are media swarms that imitate themselves (Tarde 1983), what alterities are produced? If their actions occur mainly through Facebook, does it seem possible for there to be the formation of a new commons—a commonality formed by alterities and multiplicities that are distinct, but not separated?

Thus, we can introduce a new analytical category to discuss this phenomenon, something called *altericide*, which is a term suggested by Dominique Quessada. For Quessada (2007), it has been common to think the idea of alterity as that which separates the I from the Other or as those who recognize themselves in this separation. This assertion is heir to the Western epistemological tradition founded on the idea that separability is the beginning of the comprehension of aspects of the real. Therefore, the phenomena, when analyzed according to this logic, always need to be decomposed. In other words, *analyzing* means cutting or fragmenting. Needless to say that since Plato and Aristotle the exercise of reason, or everything that is called Western rationality, rests on such a model. A real that institutes essence holds entities closed in on themselves: "In this way, separation constitutes the fundamental operative modality of Western culture: to know becomes to separate elements" (Quessada, 2013, 14). These elements come to be perceived as carriers of dualities: essence and appearance, the inside and the outside, the part and the whole, humans and objects or, even, nature and culture. Our approach is therefore formed in the negative dialectics that comes out of these dualities.

In another direction, Quessada notes that it is necessary to overcome these dualities from the perspective of post-dialectical thinking. The author reminds us that this does not aim to abolish the different or the strange, but to accept the other and the world of things as inseparable. Now, if we conceive of ourselves in this way, the other ceases to exist as before and is recognized as one who is close to and involved in a common and complementary activity. Quessada stresses that the widespread altericide, that we participate in and observe, in which we know of it whether we want it or not, is linked to a "system of a world-wide speech or to a speech empire"—a system of communication.

> The biopolitical integration on which rests the effectiveness of this empire—without the Other, once it is founded upon the consummation of itself—does not, essentially, distinguish political, economic, social and cultural issues. It turns the Other into a function played by the ontological obsolescence, in which the economy expects to become the very ontology. In this way, Being can express itself without having to resort to another. (Quessada 2007, 55)

This is, therefore, an unprecedented aspect of the crisis of the separability paradigm. We think that the Internet radicalizes such a crisis, in that it opens more sociability based on traditional molar identity principles (Deleuze and Guattari 1996). Now we are, it seems, joined by the connection and interaction of flows, which are the generators of diverse communities. Such communities, although diverse, have contradictory modulations. At the same time that they seem to be designed to avoid solidifying properties, they also close themselves off. Facebook is an example of such modulations, because I can create and/or belong to multiple communities: a grouping of friends, a network of researchers, a list of parents or relatives, etc. We poke the profiles of our friends. We "like" their photographs. We make posts, and we create associations as dyadic spheres (Sloterdijk 2003), that is, pairs of connected individuals who are, at the same time, singled out in their virtual enclosures.

3. NEW DWELLINGS: INDOOR AND OUTDOOR ARCHITECTURES

We can call this new dwelling a double movement. On one hand, we experience the indoor spatiality of the relationship between spaces in the intimacy of the house and the spaces of computers and other communicational devices that are a part of it. On the other hand, there is the outdoor spatiality that is mediated by display screens and mediatic tools, especially smartphones, TVs, Web 2.0 applications, and the street itself. In cyber life, the traditional spaces of the atom no longer have a dominant place. In this world of "foam" (Sloterdijk 2006) or "liquid"

(Bauman 2007), the fundamental ontological question is no longer "who we are" but "where we are." Just observe, for instance, a user's behavior with his/her smartphone, absorbed in an informational miniworld. We think that such behavior limits perspective. Furthermore, there appears to be a loss of laterality. In this way, the Other who interests me should fit in the palm of my hand. The loss of perspective, I would argue, can lead us to a loss of proximity and moral responsibility in the face of the Other. Such a loss is called, by Bauman (2014), *adiaphorization*. By this, he means those actions exempt from ethical evaluation, and therefore, without accountability, *a priori*, for the space of the other. Adiaphorization, therefore, makes life meaningless. The subject becomes a database, set-up as a digital market in which the individual only has moral value when consuming and revealing their data. Otherwise, they are doomed to being Balzac's Père Goriot or Colonel Chabert—Insignificant! Therefore, we must understand the actions of the online scenic masses in from a seemingly paradoxical perspective. It is with the same technical devices that approach and connect subjects that it is possible to institute this ethical and communicational gap with the other. Social networks and their affordances are true online agoras. This is a contemporary feature, but it is expressed by the hegemony of an artist capitalism or a performative and aesthetic capitalism (Lipovetsky and Serroy 2015).

We can highlight Facebook, Instagram, Twitter, and Tinder as spaces that shelter and enable this double spatiality. Also noteworthy are the groups or communities formed by WhatsApp, although we recognize that the other applications have garnered greater attention for both quantity of information and content production. If we typically question ourselves about time, we affirm that the most relevant epistemological questions today concern the question of a philosophy of space and the new forms of human sociability. Of course, we cannot respond to such questions by referring the inquiry only to geographers and cartographers. What is increasingly evident and important are the connections established by the inhabitants of these virtual, miniworld. Would Facebook and, more recently, WhatsApp not be miniaturized spaces of existence? And would portable devices not be their mobile architectures and confinements? Sloterdijk suggests that the radicalization of such confinement would be the apartment—a kind of architectural cell that explains the egoistic topology of modern individualism. It is, then, a spatial immunology that is only intensified in the cyber-conviviality of applications like Facebook and WhatsApp. A kind of indoor living that is, at the same time, outdoor, since it seems impossible to define the boundary of the domestic space, when using these communication platforms.

4. THE HETEROTOPIC SPACES

If the city is a form of unveiling (to use Heidegger's term) and man's relation to space is a topological enigma to be deciphered (Freud's concept), how can we think these issues from the perspective of heterotopia (a "strange place" or a space that is "wholly other," absolutely different . . . a counterspace)—a heterotopology.

> I say a science—which would have as its object these different spaces, these other places, these mythical and real contestations of the space in which we live. This science would study not utopias, because you need to reserve that name for what truly has no place, but the heterotopias, absolutely other spaces. (Foucault 2015, 20–21)

We hold the opinion that with the advent of the Internet and the technical sophistication of digital communication devices, we have instituted a dwelling in heterotopic miniworlds or absolutely other spaces. Tinder, Facebook, WhatsApp, Instagram, to name just a few, are its contemporary expression. Relationships become miniaturized by the inhabitants of these small polities. As if riding on the heterotopic flying carpets remembered by Foucault, inhabitants fly through the universe of computer networks to access to distant and foreign territories. How, then, is one to build counterspaces and/or heterotopic experiences as a resident of a city in which everything seems to be mediated by images, mediatic standards, and technology? In this regard, we must also consider the body as a kind of miniworld or place of spatial visibility. We can single out the proliferation of the selfies as an example. The egotistic imperative (I CAN, because I am seen) imposes itself as the domain of expression for the other. The whole of humanity involved in the desire to participate in self-presentation. In addition, there are "dickpics" and nudes exchanged between virtual lovers, the pornography sites, and the intensive consummation of the Id (I WANT, because I desire). Thus, what is the role of the superego (I CANNOT, because there are rules and social limits), when everything seems to go back to the individualization of habits and instincts? The body seems to be the absolute territory of a tyrannical and quickly moving individualization of the narcissistic desire, even by the simple act of appearing to oneself and to others, as well as when one's body becomes a destructive power, as is the case with suicide bombers. Obviously, in the heterotopic body, which is a body that acts by deviations, there are other spatialities. What, for instance, would be the body, when altered by amphetamines, alcohol, madness, passions and the immersive rituals of a religious faith? Just as there are enumerable states of being (Galeno 2005), undoubtedly there has also to be strange architectures. In this way, I can, following the examples provided by Baudelaire's Flâneur and the Passante, as well as the rebel, perceive, and build my counterspaces within the city.

5. OUTDOORS FACES—ROLEZINHOS[1]

If it is true that, with the advent of virtual communication, faces often become invisible or metamorphose themselves through the daily obsession with editing our profiles on the networks of social media, then we can also see them in physical spaces or in the public sphere. They, therefore, leave their indoor architectures to dwell in other, outdoor spaces and architectures. Currently we can, for example, witness the spectacle of actors who join Facebook but who also meet face-to-face in the streets. We could say that they are scenic masses who find a new way to express their political resistance starting from the telematic connections online. Very distant from the Luddites, who sabotaged their work by breaking machines during the Industrial Revolution in England, we now see the revolt of the Rolezinho, who confront the forces of multinational capitalism, but without issuing a list of demands. They are, it seems, driven by the power of a will to nothingness, which should not be confused with a will of nothingness. It is a kind of political metaphysics that disturbs the rationality of capital and the dialectic of the leftists and rightists—the tributaries of logic and rational explanation.

How can we classify them? They are not Black Blocs because they do not wear masks and are more diverse in both attire and behavior. Furthermore, they convene their meetings and demonstrations in the shopping centers, the Crystal Palaces of consumption, the contemporary and aseptic temples of shopping euphoria and extreme visibility.

The Rolezinhos are of the "Thumbelina generation" (Serres 2013). They use the Internet and make their mobile phones prostheses of sociability for cultivating communities involved in a fight for the will to nothingness. As middle class boys and girls, they appear as Flash mobs inside the temples of consumption that begin as quickly and then disperse.

They also differ from the typical manifestations of MPL (the Portuguese acronym for the free public transport movement in Brazil), since they seem to be representatives of the economic periphery who have risen up and advocate for a place in the sun in the world of the consumer. They want to be seen and recognized. Certainly, the governments of Lula and Dilma are responsible, in part, for the Rolezinhos, because they are poor but willing to shop. And they will, not necessarily to buy, but to get noticed collectively. They are a kind of a consumerist pack. They are *homo consumericus*, emotional consumers faithful to those brands bearing status (Lipovetsky 2007). If they are consumers, why do they frighten the capitalists or the elites? The answer is simple. For the elites, the Rolezinhos may have the resources to consume Nike clothes or Apple smartphones, but, and at the same time, they also carry with them the stigma of poverty. They are not accepted as equals. And so, they treat them not as consumers and citizens, but as *homo sacer* (Agamben 2010) arising from the margins of urban affluence. They are subjects without qualities and, therefore, socially disposable. We might wish that the manifestations of Rolezinhos were translated into some kind of concrete resistance against the capitalist monopolies of communication: TIM, CLARO, OI, VIVO.[2] We mention these because they are directly linked to telecommunications infrastructure and, therefore, the objects or tools of their mobilization—the cell phone.

According to Axel Honneth (2003), it is when there is no visibility or when there is no guarantee of the subjective and social recognition of individuals and groups that society encounters social tensions. The social demonstrations and protests that occurred in Brazil in 2013 had an axis of urban mobility, and the demonstrations of the so-called Rolezinhos illustrate such a formulation. The struggle for recognition is, therefore, an arena of conflict and tension that may result in new political configurations and new forms of social organization. In other words, it can result in significant changes in society. Of course, that will depend on the capacity of the groups involved and on intermediary organizations such as the State, so that such tensions do not result in violence. That is why, in a democracy, conflicts can result in gains in recognition, particularly for social groups claiming that their legal and symbolic rights be seen. As Sloterdijk (2004, 236) describes it, "In the city and only in the city, can it be demonstrated what it means that a figure tries not to remain hidden and stand in the center of the visible and of the notorious. Since there have been cities, appearance means: exhibition, presentation, permanent revelation. As Heidegger says: the construction of cities is a mode of revealing."

We therefore agree with Mongin (2013), the spatial configurations of the city currently challenge us to three configurations. The first Mongin terms the idealized city. Here we encounter the city as thought or idealized. That is, a city in which our bodies and spirits are inspired. It is an ideal-type that appears to be unattainable or utopian. Yet it still provides something by which to see, to act, and to think. The second configuration is the networked city in the era of globalization. With this the author intentionally refers to a reversal of the concept of a "network of cities," which was developed by individuals like Fernand Braudel, to another that he calls a "network of the city." This is the condition of urbanization encountered in the phenomena of megacities and integrated and deterritorialized markets. In this area, it is necessary to note that the city cannot reconstitute itself, as it will be a place connected to or interconnected with the events and the flows of information that it experiences. Thus, in the second mode, the dream of the idealized city does not seem to be possible to us anymore, given the excessive rationality of the markets and the logic of accelerated urbanization and mediation by urban experiences of a *vita activa*.

The third and final configuration, Mongin terms the era of post-cities or the cities of flows, or even, the era of a "generalized urbanism." Here we face a double phenomenon, says Mongin. On one hand, we have the experience of the "prevalence of" the most diverse "flows"—telecommunications networks and means of transportation. The network is, then, jutting out over the city and "orchestrating the center periphery relationship," but it still perceives itself as the continuation of territories. On the other hand, we witness the impacts of the transformations of urban spaces as places and territories under the "external pressure of the flows." This inversion, Mongin argues, is a product of Western technology and its will for control over the city that has only become exacerbated in the current era of globalization. This demonstrates that, with the advent of globalization, the territory and the place continue but expand themselves. In the post-city and the widespread urban areas, the distinction of places and a resumption of territories become explicit by privileging networks and information flows. This also explains how globalization or the "world society," as Edgar Morin advocates, was not the end of the territories, but rather meant "territorial reconfigurations," in which individuals experience the future of global cities, megacities, metropolis and megalopolis together with what we call "new economies of scale."

What is important to note in the polyphony of these three configurations is the place of the city in conflicts and tensions. It will be in the "dialogue" of conflict and tension that the interest of groups or diverse social actors can be observed, because, if one agrees with Mongin's ideas, in the current urban scenario it is no longer a matter of the classic forms of tension formulated by the Marxist paradigm of the Class Struggle that will be experienced. It will instead be a Struggle of Place. It will be here that territorializations, deterritorializations, and reterritorializations of the struggle for recognition will take place. It is also noteworthy that the place of the struggle for recognition is not located only in the physical or geographical terrain of the city. They are part of the regimes of visibility that appear in many different kinds of media. For Sloterdijk, the struggles for recognition in modernity have shifting from the traditional city space of the streets to the networks of mass media. Therefore, for him, the cultural and political struggles today are products of the so-called masses of the media. For Sloterdijk this was limited to the technology of television, but we should expand this consideration and add to it the Internet as a space for the struggle for recognition in the era of post-cities and information networks. Consequently, the Rolezinhos are mass media masses who have found a new way to express their political resistance within the social networks of the Internet. This explains the paradox that we encountered at the beginning. That is, digital communication happens in a faceless hypersphere but also enables the emergence of new facets that emerge in this new public space.

NOTES

1. Movement of young people from São Paulo's periphery, which mainly occurred in 2013. Organized over the Internet, they had their meeting points in the shopping centers and malls of the city. *The New York Times* defined them as "little scrolls."
2. Private telephone companies in Brazil.

REFERENCES

Agamben, Giorgio. 2010. *Homo Sacer. O poder soberano e vida nua I.* Belo Horizonte: Editora da UFMG.

Balzac, Honoré de. 1993. *Louis Lambert.* In: BALZAC, Honoré de. *A Comédia Humana: estudos analíticos.* Vol. 17. São Paulo: Globo.

Bauman, Zygmunt. 2007. *Vida Líquida.* Rio de Janeiro: Jorge Zahar Editores.

Bauman, Zygmunt. 2014. *Cegueira Moral. A perda da sensiblidade na Modernidade Líquida.* Rio de Janeiro: Zahar.

Deleuze, Gilles and Guattari, Félix. 1996. *Mil Platôs. Capitalismo e Esquizofrenia.* São Paulo: Editora 34, 3v.

Foucault, Michel. 2015. *O Corpo Utópico, As Heterotopias.* São Paulo: N-1 edições.

Galeno, Alex. 2005. *Antonin Artaud. A revolta de um anjo terrível.* Porto Alegre: Sulina.

Honneth, Axel. 2003. *Luta por Reconhecimento.* São Paulo: Editora 34.

Lipovetsky, Gilles and Serroy, Jean. 2015. *A estetização do mundo – Viver na era do capitalismo artista.* São Paulo: Companhia das Letras.

Lipovetsky, Gilles. 2007. *A Felicidade Paradoxal. Ensaio sobre a sociedade de Hiperconsumo.* São Paulo: Companhia das Letras.

Marceau, Félicien. 1986. *Balzac et son monde.* Collection n°108. Paris: Gallimard.

Marcondes Filho, Ciro. 2013. *O rosto e a máquina. O fenômeno da comunicação visto pelos ângulos humano, medial e tecnológico.* Nova Teoria da Comunicação. Volume 1. São Paulo: Paulus.

Marcondes Filho, Ciro. 2012. *Fascinação e Miséria da Comunicação na Cibercultura.* Porto Alegre: Editora Sulina.

Mongin, Olivier. 2013. *La Ville des flux. L'envers et l'endroit de la mondilisation urbaine.* Paris: Fayard.

Moretti, Franco. 2011. *Teoría de Redes: análisis de trama*, New Left Review, Madrid, n.68.

Quessada, Dominique. 2007. *Court Traité D'Altéricide.* Paris: Verticales.

Quessada, Dominique. 2013. *L´inseperé. Essai sur un monde sans autre.* Paris: Presses Universitair.

Rimbaud, Arthur. 1997. *Correspondência de Rimbaud.* Porto Alegre: L&PM.

Serres, Michel. *Polegarzinha.* 2013. Rio de Janeiro: Bertrand Brasil.

Sloterdijk, Peter. 2003. *Esferas I. Bolhas. Microsferologia.* Madrid: Ediciones Siruela.

Sloterdijk, Peter. 2004. *Esferas II. Globos. Macrosferologia.* Madrid: Ediciones Siruela.

Sloterdijk, Peter. 2006. *Esferas III. Espumas.* Madrid: Ediciones Siruela.

Tarde, Gabriel. 1983. *As Leis da Imitação.* Lisboa: Rés.

Chapter 7

(De)Facing Alterity in the Digital Age

"The Real Problem" in the Social Interaction of Digital Natives

Ann Hetzel Gunkel

This chapter examines research on students abroad who deploy digital communication technologies and reflects on two related theoretical problems located in such work: the question of alterity and the nature of the real in the digital age. I locate these remarks as a reflection on the purposes, procedures and products of the research carried out by the 2012–2014 Harmonia Grant project, "Negotiating Cultural Differences in the Digital Communication Era" at the Jagiellonian University in Krakow, Poland, a two-year quantitative and qualitative research project on the communicative interaction of international university students studying abroad, documented in the collection *Digital Diversities: Social Media and Intercultural Experience* (Robson and Zachara 2014). This chapter begins by asking about the philosophical assumptions of our research, the implications of our questions and the concepts being mobilized by the study.[1] My goal in these remarks is less to interpret the data set that has been gathered but rather, to complicate the very premises we deploy in framing research of this type. I want, in the most positive way, to make our work suspect [from the Latin sub- + *specere*, to look at from below, from underneath, from the foundation, as it were] so that in making our work suspect, we can trace in the project, the underlying assumptions that shape and in many ways determine the meaning and significance of our study. Specifically, I interrogate three areas for questioning; all three are interdependent and implicate each other at a fundamental level. They are: (1) the use of the term *digital natives*, (2) the limits of qualitative methodology in studying digital natives, and (3) the concepts of the real and the Other deployed by digital natives.

1. CONSIDERING DIGITAL NATIVES

The first is the seemingly self-evident terminology describing our subject group: digital natives. What does this term actually mean? We use it effortlessly to indicate some qualities of the interview subjects. But who exactly are digital natives? Who counts as a digital native? Whence this term? And perhaps most importantly, what are the consequences of deploying a marketing term with a distinct colonial and imperial legacy? How does the implicit anthropological formation of the object of study determine and shape that study even before it begins? As a Cultural Studies scholar, I am deeply suspect of framing a research subject in the language of a marketing scheme. And quite frankly, even more worried about adopting the colonialism of an anthropological gesture toward the "native population" of cyberspace, re-deploying this nineteenth-century term devoid of its political and imperial context.

When positing or deploying a "native identity" formation such as digital natives, it might behoove us to consider extant critiques of ethnic identity scholarship. "Images . . . [of ethnicity] . . . purveyed by the mass media are neither just the compilation of folk ideas nor the popularization of scholarly findings, but also reflections of the needs of capital and the state. This material link is most easily seen in the advertising media, which not only describe products but also manipulate images of women, men, and children so as to define them as individuals needing those commodities" (di Leonardo 1984, 178). We see that material link in the ways the term *digital natives* has been deployed as a marketing category, with the proliferation of hundreds of articles bearing titles such as "Digital Natives: Six Ways Marketers Can Engage Millennials" (Sparks 2013). The term *digital natives*—despite its questionable descriptive capabilities regarding an entire global generation—is fundamentally a marketing category, aimed at selling even more effectively to the consumers who have a personal relationship with Beanie Babies, Tamagotchi and Slap bracelets (Sparks 2013). Conversations in boardrooms and blogs resonate with seminars such as "Cracking Today's Digital Natives: 5 Things to Keep in Mind When Marketing to Millennials" found on the Word of Mouth Marketing Association website which advises, "More than any other generation, millennials value relationships with brands that are authentic and have a one-on-one feel" (Jordan 2013). We need to be skeptical and wary of mobilizing marketing categories representative of what Naomi Klein (2000) calls the triumph of identity marketing as descriptors for academic research. Digital natives are in fact, the prized target of so-called "cool hunters" scholars with academic training in the social sciences and humanities, particular anthropology, hired to take on "ethnographic field work" on the demographics of coolness for corporate purposes (Klein 2000).

Moving from marketing to the field of higher education, educational theorist and game designer Marc Prensky claims credit for popularizing the term *digital native* in his article, "Digital Natives Digital Immigrants."[2]

> What should we call these "new" students of today? Some refer to them as the N-[for Net]-gen or D-[for digital]-gen. But the most useful designation I have found for them is Digital Natives. Our students today are all "native speakers" of the digital language of computers, video games and the Internet.
>
> So what does that make the rest of us? Those of us who were not born into the digital world but have, at some later point in our lives, become fascinated by and adopted many or most aspects of the new technology are, and always will be compared to them, Digital Immigrants. (Prensky 2001, 1)

While asking some crucial questions about the pedagogical methods used to engage students growing up in a different world from many of their teachers, he nonetheless deploys a deeply problematic quasi-anthropological formation to designate generational differences of technological enculturation, using terms such as the "digital immigrant accent." Prensky's tone (2001) is earnest because he is addressing the important question of twenty-first-century learning styles and the efficacy of outdated pedagogies. However, once again, in service of that serious question he unproblematically deploys an ethnocentric anthropological formulation that—in its wildfire acceptance, especially in popular media—has shaped our assumptions about these "digital natives." He pleads his case:

> It's very serious, because the single biggest problem facing education today is that our Digital Immigrant instructors, who speak an outdated language (that of the pre-digital age), are struggling to teach a population that speaks an entirely new language. This is obvious to the Digital Natives—school often feels pretty much as if we've brought in a population of heavily accented, unintelligible foreigners to lecture them. They

> often can't understand what the Immigrants are saying. What does "dial" a number mean, anyway? (Prensky 2001, 2)

While his educational and pedagogical concerns may be quite valid, his plea for the pedagogy of gaming is in fact buried in the baggage of colonial and nativist discourse that shapes claims made about the "new generation."

On the most practical level, critics of the term have noted it suggests a familiarity with technology that not all children have, effectively ignoring the complex political economic contexts within which one has access to technology and the simple facts of unequal access globally (Holton 2010; Jones et al. 2010; Kennedy et al. 2010; Jones and Shao 2011; McKenzie 2013). Global statistics concerning the digital divide estimate that perhaps merely 8 to 10 percent of the world's population have Internet access.

> Writing in the British Journal of Education Technology in 2008, a group of academics led by Sue Bennett of the University of Wollongong set out to debunk the whole idea of digital natives, arguing that there may be "as much variation within the digital native generation as between the generations." They caution that the idea of a new generation that learns in a different way might actually be counterproductive in education, because such sweeping generalisations "fail to recognise cognitive differences in young people of different ages, and variation within age groups." (*The Economist*)

Although the Harmonia study interviewed international students from all continents and eight languages, one cannot presume that this group is therefore globally representative. In fact, while several of the students in the sample come from developing regions, these young people are not representative of the global population. In fact, "international students comprise a highly populated sojourning group with some specific characteristics that make their experience different from other migrating groups such as guest workers or refugees" (Berry and Sam 1997, 92). As Cemalcilar, Falbo, and Stapleton point out, "They are a more homogenous group in that they are typically young and well educated. In general, they arrive in the host country pre-trained in the host language and prepared to adjust to the host culture" (Berry and Sam 1997, 92). Because of their special status, one much more privileged than many migrants, refugees, and guest workers, "keeping in touch with their own culture and society and maintaining existing relationships may be more of a need for student groups, compared to more permanently settled and established groups such as ethnic groups or immigrants" (Berry et al. 1989, 135–186). The research data completely supports these earlier studies with all but one student reporting their primary use of computer-mediated communication is contacting friends and family "back home." The survey respondents quite clearly echo the existing research in the field, in that they pointed out not only the primary function of technology for contacting home but also their lack of using that exact same technology (i.e., Facebook) for contacting people in the host country. As a foreign student studying in Poland commented, "I rarely use it to keep in touch with friends here in Cracow." When considering a study of international university students abroad, one might presume that their digital interactions might engage with alterity in some form. The Harmonia study suggests that this is *not* the case. Instead of engaging with "the Other" through computer-mediated communication, it seems that our sample used that technology primarily to engage with the same, the familiar and the comfortable.

A second interesting feature of the relative privilege of the group is the almost universal practice in our sample of posting travel photos. While the respondents varied in the level of usage and opinions about digital technologies, almost all of them reported engaging in one practice: posting travel photos. Perhaps the ultimate visual icon of modern cosmopolitan identity, few artifacts measure up—in both ontological weight and sheer surface gloss—to the photograph. Susan Sontag (1973, 71) claimed that, "As photographs give people an imaginary possession of a past that is unreal, they also help people to take possession of space in which they are insecure." Rather than assimilating or enculturating, it seems that a predominant digital practice of international students is the reproduction of the tourist gaze and its attendant practices. What Urry and Larson (2001) call "the tourist gaze" is a mode of seeing and representation that regulates the relationship with the tourist environment, demarcating the Other and identifying the out-of-the-ordinary. It elucidates the relationship between tourism and embodiment and elaborates on the connections between mobility as a mark of modern and postmodern experience and the attraction of tourism as a lifestyle choice.

The effects of tourism on "natives," particularly the commoditization of culture, are increasingly subject to study. From the early scholarship of Veblen (2009) on the leisure class to contemporary analyses of tourism, scholars have asked about the economic and imperialist context of travel extending their analyses to the commodification of "exotic" locales by foreign tourists (Enloe 2000). We might want to explore how digital natives in their touristic practices construct their participation in modernity and their status as modern subjects on Facebook. "To be a tourist is one of the characteristics of 'modern' experience. Not to 'go away'; is like not possessing a car or a nice house. It has become a marker of status on modern societies" (Urry and Larsen 2001, 3). Numerous texts have explored how "the camera and the tourism are two of the uniquely modern ways of defining reality" (Horne 1984, 21). In fact, the concept of the gaze, as a constitutive part of modernity highlights "that looking is a learned ability and the pure and innocent eye is a myth" (Urry and Larsen 2001, 1). The classifications made by the tourist gaze occur within an economy of relations, producing what Said called "imaginative geographies" (Said 1995, 49–73).

Digital photography has expanded the role of the tourist gaze in the space of social media. "Users of Facebook have uploaded more than 10 billion photographs, with the number increasing by an astonishing 700 million each month" (Urry and Larsen 2001, 185). Most of the research sample respondents not only post travel photos as a matter of course, but almost all of them mention posting landscape photographs. "The tourist gaze is directed to features of landscape and townscape which separate them off from everyday experience. Such aspects are viewed because they are taken to be in some sense out of the ordinary" (Urry and Larsen 2001, 3). The almost universal practice of posting touristic landscape photos in the Harmonia sample points to not only the privileged status of international students as modern subjects but reinforces the questions raised about mobility, postmodernity, representation and the legacies of colonialism.

Indeed, we might ask, following bell hooks (1992), whether or not digital technologies are deployed not to enable an encounter with Alterity, but rather to consume alterity as a kind of manageable commodity. In her oft-quoted essay, "Eating the Other: Desire and Resistance," hooks (1992, 21) notes that "The commodification of Otherness has been so successful because it is offered as a new delight, more intense, more satisfying than normal ways of doing and feeling. Within commodity culture, ethnicity becomes spice, seasoning that can liven up the dull dish that is mainstream white culture." The students in the

Harmonia survey posted photos, especially on Facebook, but did not use that format to actually interact with local people. Instead, they posted touristic markers, icons of "otherness" that indicated their encounter with the desired Other—"Look! I've seen the Other"—as a "way to make themselves over," to leave behind "innocence and enter the world of experience" (23). Deploying a series of markers meant to indicate an exotic locale, ethnic otherness, and a more worldly posture, the Facebook poster indicates that they are now on the world stage. But "the acknowledged Other must assume recognizable form," as hooks (23) reminds us. Imperialist nostalgia operates through a tourist gaze framed around fantasies of the primitive, the "ethnic," the "authentic" and the Other.

We might ask, following Sontag, whether digital natives via their photo posting practices are indeed enacting a "chronic voyeuristic relation" to the world around them. Or perhaps, following Roland Barthes (1981), we might examine photography's tendency to naturalize highly structured meanings. In the texts of *Mythologies*, Barthes (2009) explores Myth (including photographic images) as a type of speech; delimiting how the process of mythologization brings a truth claim to socially constructed notions, narratives, and assumptions. The rhetorical power of photography, made in the now-moment by digital photography and instantaneous posting, is grounded in the ability to naturalize, to make innocent its cultural messages and connotations. "Photographs appear to be not statements about the world but pieces of it, even miniature slices of reality, without revealing its constructed nature or its ideological content" (Urry and Larsen 2001, 168). The claim made upon the real made by photography might prove a most fruitful path for analyses of social media data in that it connects these digital practices directly to the question of the real (to which I will turn in Section 3).

While we must remain cautious about the effectiveness of categorization in the term *digital native*, we need also to consider the epistemological assumptions made about that group. As poster drkhturner notes on the Digital Natives debate site of discusscafe (2013), "If digital natives are people who were immersed in particular digital technologies during their formative years, then yes, they exist. The challenge is that they may see these same technologies as "natural"—and that they may not be as skilled, self-aware, or critical in their use of the technologies as we might assume they would be." In other words, even if the category of digital natives is valid, it does not follow that said digital natives have a privileged understanding of technology, the parameters of which have been naturalized. This observation leads to my comments about the relationship of digital natives to the concept of the real, to which I turn in section 3.

There are a flood of debates, discussions, wikis, and blogs devoted to the argument about digital natives. However, in those numerous sites, what is debated is whether or not the term is valid and whether or not it can be used to describe an entire generation unproblematically. While this is a worthwhile inquiry, it seems to me that the critiques miss something more fundamental: namely, what does it mean to describe an entire generation in terminology derived from colonialism? To redeploy in a supposed post-racial era, the nomenclature of natives and immigrants? Micaela di Leonardo's (1998, 38) critique of the anthropological stance of going native notes, "The Other is terminally Orientalized—a proven inferior who must be forced to cooperate in studying his or her own present or past, an exotic individual who, in the aggregate, can provide the *mise-en-scène* for an infinite series of dramas of Western selfhood. Anthropologists participate in this 'colonial chic' or imperialist nostalgia."

On this point, previous scholarship examines the link between formulations of cyberspace as the new frontier and the colonial project. Following Heidegger, we observe that philosophical terminology cannot be divorced from the history of the use of that terminology, and thus we cannot avoid confronting questions of language and meaning.

> First, metaphors are always more than mere words. They are mechanisms of real social and political hegemony that have the capacity to determine the current and future shape of what they merely seem to designate. As a result of this, current and future configurations of cyberspace will be determined not only through innovations in hardware and software, but also, and perhaps more so, through the various metaphors that have been circulated and are employed to describe their significance. . . . Because cyberspace has already been submitted to a kind of colonization through the metaphors of the new world and the electronic frontier, its decolonization is a task that, if it ever transpires, must take place in and by engaging the material and legacy of these particular rhetorical configurations. (Gunkel 2001, 51)

In the case of the term *digital natives*, who appear in all research to represent a privileged sample of education and mobility, it is particularly notable that this term appears in the supposedly post-racial era where some commentators have wondered why first world scholars have jettisoned the concept of subjectivity at precisely the moment when third world and native peoples were claiming it. This requires, at the very least, the decolonization of our categories of research.

2. ON THEORY AND METHOD: QUALITATIVE STUDY AND SELF-REPORTING

The attendant problem of studying the "native population" of digital culture is the corollary assumption that the natives have a privileged, that is to say unmediated access to their own conditions of existence. This presumption leads us to question the underlying principles embedded in any qualitative methodology, especially those that involve self-reporting. This is not to say that the research is not useful or productive, nor is it to reject that methodology. But I want to take the step back through a consideration of the limitations and complications of certain implicit theoretical moves so as to clarify and qualify what it is we can and cannot claim. As David Gunkel (2014) points out, self-reporting has the distinct advantage that it can, unlike any other form of data collection, provide access to participant's thoughts, motivations, emotions, and gratifications. The main disadvantage, of course, has to do with the validity of reported data. This not only involves deception on the part of participants (deliberate or otherwise) but also "leading questions" from the interviewers, which can influence how one responds. Furthermore, self-reporting is based on an essentially modernist assumption that subjects are transparent to themselves, know what they do, and can reasonably explain why they do it (Gunkel 2014, 139). This presumed self-transparent subject is further weighted with assumptions about the proximity of digital natives to their native realm. The modernist view of the subject translates directly into/is based upon the modernist, instrumentalist view of the real.

The term *digital native* seems to presume some more immediate, more automatic, perhaps more intuitive understanding of technology. Our subjects, we presume, are somehow native inhabitants of digital culture. Further, it is generally presumed that this immediacy must mean a more sophisticated understanding of how that technology works in the world and what that means. My analysis claims that no such superior understanding of the virtual is demonstrated by digital natives, who deploy the very same rhetorical and conceptual frameworks that dominate mainstream commentary, academic studies of communication, and the viewpoints of non-digital natives. Indeed, I want to argue that digital natives express views of the virtual that are closely aligned with commentators and philosophers from ancient through modern times—all of whom predate modern computing. In short, digital natives seem to have no advantage or privileged understanding of the metaphysics underlying the technology with which they have greater familiarity.

That interrogation of the limits of methodology points toward an even more fundamental formulation and the third concept I want to examine: the modernist, instrumentalist view of the technology that animates our interview subjects connects directly to the modernist view of the real underlying their assertions. Our data can tell us about the behaviors and practices of international students in online space, but we cannot—via qualitative interviews, no matter how rich—make claims about the nature of real social interaction as opposed to virtual interaction. Our subjects do, however, make those claims as a matter of course. What we can learn from them is precisely how their rhetoric of the real informs their understanding of their own actions online and offline and furthermore point to the seeming hegemony of that philosophical concept.

3. CONCERNING THE REAL

What is clear from the Harmonia data is that digital natives certainly mobilize the language and rhetoric of the real; in fact, several of them comment on the difference between real and virtual life as though this was a self-evident distinction. This is very informative, not because they are "reporting on" the real/virtual opposition, but because it is an organizing concept deployed by almost everyone in the interviews. We might argue that digital natives have a profound investment in the Platonic metaphysics of reality because while virtually all subjects responded by mobilizing the rhetoric of the real at some point in their interview, the interview questions never used that language. Even presuming a caution on the part of our study against leading the subjects to such comments, they emerged seemingly unbidden.

> First, everything depends on how we define and operationalize the concept of the real. Even though online role playing games, social networks, and other forms of avatar-based CMC are often considered to be merely a matter of entertainment, they are involved in serious debates about and meditations on fundamental aspects of metaphysics. And in these situations there appears to be, as there are in many facets of computing, a default setting. This default has been programmed and is controlled by Platonism, which institutes a distinction between the real thing and its phenomenal appearances. In computer-mediated interaction, like online role-playing games and immersive social environments, this Platonic decision is particularly manifest in the discussions and debates surrounding avatar identity and the seemingly indisputable fact that what appears in the space of the virtual world are manipulated representations of real human users, who may themselves be entire different from how they appear in the computer-generated environment. As long as our research endeavors remain within and proceed according to this Platonic formulation, which as a default setting is often operative without having to select or specify it, we already know what questions matter, what evidence will count as appropriate, and what outcomes will be acceptable. (Gunkel 2010, 127)

I want to ask whether digital natives have even begun to problematize the notion of the real—firmly in place since Plato and through Descartes—that animates many of their conclusions and observations. Digital natives—whoever they may be and however tech savvy—don't seem to have problematized the underlying metaphysical concepts that the virtual world might have or should have called into question. The students in our sample may be "digital natives" but they have roughly the same theoretical perspective on the virtual world as Descartes, exhibiting and deploying what might be called Descartes 2.0 (Gunkel and Hetzel Gunkel 2009, 104–127).

I will argue that digital natives have almost no advantage in interrupting or questioning the parameters of reality presumed by rationalist or operationalist concepts of technology. While the Harmonia survey population is quite young (average age of 25.6), they nonetheless embody varying ancient and medieval notions about technology and metaphysics, with one participant claiming a neo-luddite view that, "A world without an Internet would be better" because computer-mediated technologies are "artificial and superficial." In other words, the underlying concept of the real being mobilized by these young people is one straight from Platonic metaphysics where the real is coextant with an originary entity and all representations are removed to varying degrees from the source. Artifice or superfice is presumed to be derivative, and thus, inauthentic. It is fascinating to see the Platonic schema of appearances and reality mobilized by twenty-first century digital natives, who—following Plato, in a reading straight from Book 10 of *Republic*—worry about the inauthenticity of appearances in the digital realm. As one interviewee noted, "People are not being real on Facebook." As Boellstorf (2008, 119) points out, "There is a gap between virtual and actual self . . . and a broadly shared cultural assumption that virtual selfhood is not identical to actual selfhood." What our respondent seems to be suggesting by commenting that "people are not being real on Facebook" is that new technologies allow for a performance that should be suspect. What this doesn't take into account, of course, is whether or not subjectivity is always already a performative.

While most all of the respondents regularly use computer-mediated communication, most of them find it suspicious and possibly lacking in metaphysical authenticity. One student commented that he/she prefers "pre-Internet experience" and yet another, "I very much prefer my social interactions with people before Facebook. Or even phone." The dual suspicion of Facebook and the telephone is very telling. A quick glance at the history of technology reminds us that the new invention of the telephone was met with precisely the same rhetorical critiques now being leveled at CMCs. In fact, the phone was presumed to create the possibility for deception in social interactions, since you could not see the person you were talking to. An historical perspective on the fear of emergent technologies is very informative for contextualizing digital culture. "With the advent of the telephone and other new media came relatively sudden and largely unanticipated possibilities of mixing heterogeneous social worlds—a useful opportunity for some, a dreadful intrusion for others. New media took social risks by permitting outsiders to cross boundaries of race, gender and class without penalty" (Marvin 1988, 107). In short, the telephone as a new media form proposed the crossing of socially established boundaries of class, race, gender and nation, such that the "wrong" kind of person just might call you on the phone. And this threat is very much the same threat implied in the concern over cyber communication where you might not know exactly whom you're talking to. On one level, technophobia is a fear of miscegenation. "The telephone and other new media introduced a permeable boundary at the vital center of class and family, where innovative experiments could take place in all social relations, from crime to courtship" (Marvin 1988, 108). How strange it is that the social upheaval of the telephone mirrors almost exactly the social upheavals linked to the Internet, in particular, worries about social relations from "crime to courtship," in our time, cybercrime and online dating. The social fears around these new developments produced social policies attempting to limit "illegitimate" access to telephones (Marvin 1988, 104–105). Even more immediately, the fear of new technologies is so powerful that our very well-being and even sanity is said to be at stake. Newspaper accounts in the 1880s and 1890s credit the telephone with causing insanity, in one case reporting "that the telephone had driven a Cincinnati citizen insane" (Marvin 1988, 187).

This concern over the "reality" of the interlocutor was applied to telephone conversation and now Internet avatars using almost the exact same language and metaphysics. This fear of deceptive interaction privileges the face-to-face as the only—or at least, most authentic—means of human communication. This, too, is directly reflective of the Platonic schema of the *Phaedrus* which famously poses the new technology of writing (seen as an inferior, and thus, inaccurate representation of speech) as a threat to the unity and authenticity of the voice (Plato 1987). As Walter Ong (1982, 79) notes, "Most persons are surprised, and many distressed, to learn that essentially the same objections commonly urged today against computers were urged by Plato in the Phaedrus (274–277) and in the Seventh Letter against writing."

Once again our digital natives seem to have bought into Platonic metaphysics wholesale—at least when it comes to their assessment of technology. This is not particularly odd or unexpected, for as Heidegger certainly reminds, our very understanding of the question concerning technology emerges from the Platonic explication of *techne*. "Plato was thinking of writing as an external, alien technology, as many people today think of the computer. Because we have by today so deeply interiorized writing, made it so much a part of ourselves, as Plato's age had not yet made it fully a part of itself (Havelock 1963, 20–60), we find it difficult to consider writing to be a technology as we commonly assume printing and the computer to be" (Ong 1982, 81). The dangers attributed to that technological invention—perhaps one of the most world-changing technological revolutions—are the template by which digital technologies are framed, understood and articulated.

This operationalist view of technology, condemning technology to either the status of a mere tool or in a more morally weighted view, the means of communicative deception, stems from a fundamentally problematic definition of communication itself, one that until only recently had gone uninterrogated even—and perhaps especially—within the field of Communication Studies (Carey 1992, Chang 1996, Gunkel 2001). There is simply no way that these presumptions of communication can remain undisturbed after phenomenology and post-structuralism. The presumption of the dominant sender-receiver model of communication depends on the notion of the self-aware, self-present subject. As Briankle Chang (1996, 181) explains, such an understanding of the communicative subject is a theoretical fiction, a Cartesian-based subject "that misses the fact that individuals are constituted as functioning communicators only insofar as they participate in communication, only insofar as they are positioned as sender or receiver differentially according to the medium and the context of a particular communicative event."

An important consequence factoring in this vital critique of communication theory is that it actually attends, most seriously, to the face-to-face that our interview subjects seem to want to privilege, because it understands the communicative context and the effect of the other—in a Levinasian sense—as the *a priori* foundation for communicative interaction. What the contemporary communicative critique offers is the possibility of restoring the face-to-face to a genuine rather than fictional function. It would be one constitutive event of communication among others instead of *the* constitutive event. This would allow for a kind of Levinasian reading in that it understands the communicative context and the effect of the other as prior to any kind of messaging. There is no *a priori* self-possessed subject totality, pre-given and decontextualized, but rather an addresser who is first addressed by the other. Michel Foucault (1981, 48–78) observes that the image of communication as infinite, free exchange of discourse represents one of the "great myths of European culture." Heidegger's *Being and Time* problematized these assumptions, which were lodged in a prior concept of presence. "Communication is never anything like a conveying of experiences, such as opinions or wishes, from the interior of one subject into the interior of another. *Dasein*-with is already essentially manifest in a co-state-of-mind and co-understanding" (Heidegger 1962, 205).

Several of the interview subjects in the Negotiating Cultural Differences study commented that while they use, enjoy, and waste time on Facebook—with some reporting usage "all day," "24/7," and "all the time"—they nonetheless accord it an inferior metaphysical status, commenting that it is "less personal than face-to-face" communication. One interview subject was asked about how digital media helps with intercultural communication and was careful to clarify, "I think the most important for me is contact live, face-to-face." This fantasy of an unmediated communicative act is just that, fantastic. Our respondents presume that whatever communicative digital practices they embody or especially, avoid are in some way inferior, less direct, and more "mediated" than the face-to-face.

Within the study data, it was extremely uncommon for respondents to complicate this metaphysics of the real, with only one digital native commenting, "I think our online social life is an important part of daily life. Not a replacement for real life but another dimension of it." This was perhaps the only incursion in the entire sample even remotely problematizing the notion of the real. The real problem (for our respondents as well as our research) is not that investigators of computer-mediated social interaction have used one theory of the real or another. The problem is that researchers have more often than not utilized theory without explicitly recognizing which one or considering why one comes to be employed as opposed to another.

The international students in the Harmonia study—representing many languages and cultures—seem to indicate that global youth culture is certainly homogenous enough to reproduce a specific metaphysical understanding of the real that is not specific to native country or native language, but remains firmly lodged in a Western, metaphysical conceptual nexus operationalized in our formation and understanding of technology's function. Those same conceptual parameters traced out in Heidegger's "Question Concerning Technology" (1977) seem to apply to this global sample. Why would a diverse international group all mobilize a singular conceptual formation? The predominance of a Platonic metaphysical formulation is not surprising in a group of international students who hail from around the globe when one looks at the existing research on international students. That brings us full circle back to the issue of privilege *vis-à-vis* the relative status of international students with respect to other global peoples on the move. for example, migrant workers, refugees, and so on. The digital natives of the Harmonia sample are relatively well educated and thus, not representative of global diversity of position and status. They are representative, however, of the ubiquitous naturalization of this specific formation of the concept of technology.

Unlike recent scholarship on virtuality that complicates this metaphysical inheritance, the students in the Harmonia sample quite clearly mobilized a single set of assumptions about the nature of the real. For contemporary theorists such as Žižek, "The real is already a virtual construct, and the difference between the real and the virtual turns out to be much more complicated and interesting" (Gunkel 2010, 46). As much of that recent scholarship points out, "It is not that virtual worlds borrowed assumptions from real life: virtual worlds show us how, under our very noses, our 'real' lives have been 'virtual' all along" (Boellstorf 2008, 5). (See also Baudrillard 1983; Taylor 1987; Zizek 2006; Gunkel 2007). In his essay on the 1999 film *The Matrix*, Žižek (2002, 240) asks about the pervasive nature of this belief, replayed over and again in our entertainment narratives. Whether it be *The Matrix* or *The Truman Show* (1998), there is a persistent, almost pre-modern notion that we are arriving at the end of the "real" universe. In our consumerist paradise, we begin to suspect that our world is a fake, a spectacle staged to convince us that we live in a real world.

As Žižek points out, this is ideology at its purest. He asks, "What if ideology resides in the very belief that outside the closure of the finite universe, there is some 'true reality' to be entered?" (Žižek 2002, 240). The nearly universal responses of the digital natives in the Harmonia study indicate that they share this vision; they must, therefore, comment upon and warn against the inauthenticity and fakeness of their own digital practices. With Žižek, I would then argue that their protests are mobilized so as to mark off the fake, their move to seemingly break out of ideology is the resumption of ideology at its purest. Why are the students in our study frantically reminding us of the un-reality of their own practices? I worry about the hegemony of common sense, a shared belief that renders digital behavior as suspect. "Contemporary experience again and again confronts us with situations in which we are compelled to take note of how our sense of reality and normal attitude toward it is grounded in a symbolic fiction," what Žižek calls "the big Other" (Žižek 2002, 249). Our deployment of "the real" via frantic attempts to discount our digital practices attempts to overdetermine the Other. "And is the struggle for hegemony not precisely the struggle for how this zero institution will be overdetermined, colored by some particular signification?" (Žižek 2002, 253). Žižek (2002, 246) states that the conservative view of reality ultimately put forth by *The Matrix* is "not radical enough." "The real is not the 'true reality' behind the virtual simulation, but the void which makes reality incomplete or inconsistent, and the function of every symbolic Matrix is to conceal this inconsistency" (246). The (ultimately conservative) thesis of *The Matrix* is, according to Žižek, that there has to be a Matrix, because things are not right. The big Other is "externalized in the really existing Mega-Computer"; thus, "the problem with the film is that it is not "crazy" enough, because it supposes another "real" reality behind our everyday reality sustained by the Matrix" (Žižek 2002, 245). This recuperative vision, a kind of paranoia about the real is precisely displayed by the respondents in the Harmonia study, who insist, again and again, that their virtual selves are not real. The effort to continually discount the fakery of our own digital practices is an effort to overdetermine the place and meaning of the inconsistency of the real, to domesticate the Other, to reign in any attempt by Alterity to break through.

It is genuinely fascinating to report from our data that so-called digital natives make frequent use of social practices of digital communication which they simultaneously suspect, according those practices less authenticity and even less reality. This betrays the struggle over the real and our dis-ease with Alterity. It is absolutely interesting that in the context of our research, our so-called native informants know just as little about the territory as we do.

NOTES

1. The research project, "Negotiating Cultural Differences in the Digital Communication Era," was sponsored by the Harmonia Grant of the Polish National Science Center/*Narodowe Centrum Nauki*. The research team consisted of Dr. Garry Robson, Dr. Małgorzata Zachara, Dr. Agnieszka Stasiewicz-Bieńkowska (Jagiellonian University, Poland), Dr. David Gunkel (Northern Illinois University, USA), and Dr. Ann Hetzel Gunkel (Columbia College Chicago, USA). A summary of the research project is provided by Gunkel (2014). Findings from the project as well as papers selected from the resulting academic conference are published in Robson and Zachara (2014).

2. In his blog, Marc Prensky (2006) qualifies his claim to originating the term Digital Natives: "Bottom line: I am the person who should get the credit

for popularizing—not for being the "first to think up,"—the native/immigrant distinction, and I should get credit as well, until an earlier citation arises, for adding the descriptor "digital." This is, of course, somewhat like, as Jerry Michalski points out, Marconi getting credit for the radio that Tesla thought up first, or Bell for the telephone thought up first by Elisha Grey and Lars Ericsson."

REFERENCES

Barthes, Roland. 1981. *Camera Lucida*. New York: Hill and Wang.

Barthes, Roland. 2009. *Mythologies*. Translated by Annette Lavers. London: Vintage.

Baudrillard, Jean. 1983. *Simulations*. Translated by Paul Foss, Paul Patton and Philip Beitchman. New York: Semiotext(e), Inc.

Berry, John W., and David Lackland Sam. 1997. "Acculturation and adaptation." In *Handbook of cross-cultural psychology*. Edited by John W. Berry, M. H. Segall and Cigdem Kagitcibasi. Boston: Allyn & Bacon, 292–326.

Berry, John W., Uichol Kim, Steven Power, Marta Young, and Merridee Bujaki. 1989. "Acculturation studies in plural societies." *Applied Psychology: An International Review* 38: 135–186.

Boellstorff, Tom. 2008. *Coming of Age in Second Life: An Anthropologist Explores the Virtually Human*. Princeton, NJ: Princeton University Press.

Carey, James W. 1992. *Communication as Culture. Essays on Media and Society*. New York: Routledge.

Cemalcilar, Zeynep, Toni Falbo and Laura M. Stapleton. 2005. "Cyber communication: A new opportunity for international students' adaptation?" *International Journal of Intercultural Relations* 29: 91–110.

Chang, Briankle G. 1996. *Deconstructing Communication*. Minnesota: University of Minnesota Press.

di Leonardo, Micaela. 1984. *The Varieties of Ethnic Experience*. Ithaca: Cornell University Press.

di Leonardo, Micaela. 1998. *Exotics at Home. Anthropologists, Others, American Modernity*. Chicago: University of Chicago Press.

Digital Natives Debate. Accessed 16 September 2013. http://www.discusscafe.com/drkhturner/digital-natives

Enloe, Cynthia. 2000. *Bananas, Beaches and Bases: Making Feminist Sense of International Politics*. Berkeley: University of California Press.

Foucault, Michel. 1981. *The Order of Discourse*. Translated by Ian McLeod. *Untying the text: A post-structuralist reader*. Edited by Robert Young. Boston: Routledge, 48–78.

Gunkel, David J. 2014. "Negotiating Cultural Difference in the Digital Communication Era." In *Digital Diversities: Social Media and Intercultural*

Experience. Edited by Robson, Garry and Małgorzata Zachara. London Cambridge Scholars Press, 120–143.

Gunkel, David J. 2010. "The real problem: avatars, metaphysics, and online social interaction." *New Media & Society* 12(1): 127–141. Accessed 16 September 2013. http://nms.sagepub.com/cgi/content/abstract/12/1/127.

Gunkel, David J. 2007. *Thinking Otherwise: Philosophy, Communication Technology*. West Lafayette: Purdue University Press.

Gunkel, David J. 2001. *Hacking Cyberspace*. Boulder, CO: Westview Press.

Gunkel, David J., and Ann Hetzel Gunkel. 1997. "Virtual geographies: The new worlds of cyberspace." *Critical Studies in Mass Communication* 14(2): 123–137.

Gunkel, David J., and Ann Hetzel Gunkel. 2009. "Terra Nova 2.0—The new Worlds of MMORPGs." *Critical Studies in Media Communication* 26(2): 104–127.

Heidegger, Martin. 1962. *Being and Time*. Translated by John Macquarrie and Edward Robinson. New York: Harper & Row.

Heidegger, Martin. 1977. *The Question Concerning Technology and Other Essays*. Translated by William Lovitt. New York: Harper Torchbooks.

hooks, bell. 1992. *Black Looks: Race and Representation*. Boston: South End Press.

Holton, Doug. 2010. EdTechDev., Accessed 12 May 2010. http://edtechdev.wordpress.com/2010/03/19/the-digital-natives-digital-immigrants-distinction-is-dead-or-at-least-dying/

Horne, Donald. 1984. *The Great Museum*. London: Pluto Press.

Jones, Chris, Ruslan Ramanau, Simon Cross and Graham Healing. 2010. "Net generation or digital natives: Is there a distinct new generation entering university?" *Computers & Education* 54(3): 722–732.

Jones, Chris and Binhui Shao. 2011. "The net generation and digital natives: implications for higher education." Higher Education Academy, York. Accessed 16 September 2013. http://oro.open.ac.uk/30014/

Jordan, Allison. 2013. "Cracking Today's Digital Natives: 5 Things to Keep in Mind When Marketing to Millennials." WOMMA. The Word of Mouth Marketing Association Blog. 26 June 2013. Accessed 16 September 2013. http://www.womma.org/blog/2013/06/cracking-todays-digital-natives-5-things-to-keep-in-mind-when-marketing-to-millennials

kennedy, Gregor, Terry Judd, Barney Dalgarno, and Jenny Waycott. 2010. "Beyond natives and immigrants: Exploring types of net generation students." *Journal of Computer Assisted Learning* 26(5): 332–343.

Klein, Naomi. 2000. *No Logo*. New York: Picador.

Kong, Ying. 2009. "Acculturation in the Age of New Media." Paper presented at the annual meeting of the International Communication Association. New York, NY. 25 May 2009. Accessed 10 September 2013. http://www.allacademic.com/meta/p13763_index.html

Marvin, Carolyn. 1998. *When Old Technologies Were New*. Oxford: Oxford University Press.

McKenzie, Jamie. 2013. "Digital nativism, digital delusions, and digital deprivation." Accessed 16 September 2013. http://www.fno.org/nov07/nativism.html

Ong, Walter. 1982. *Orality & Literacy. The technologizing of the Word.* New York: Routledge.

Plato. *Phaedrus*. 1987. Translated by Harold North Fowler. Cambridge, MA: Harvard University Press.

Prensky, Marc. 2001. "Digital Natives Digital Immigrants." In *On the Horizon.* MCB University Press, 1–6.

Prensky, Marc. 2006. "Digital Natives, Digital Immigrants: Origins of Terms." Marc Prensky's Weblog. June 12. Accessed 16 September 2013. http://www.marcprensky.com/blog/archives/000045.html

Robson, Garry and Małgorzata Zachara. 2014. *Digital Diversities: Social Media and Intercultural Experience*. London: Cambridge Scholars Press.

Said, Edward. 1995. *Orientalism: Western Conceptions of the Orient.* Harmondsworth: Penguin.

Sontag, Susan. 1973. *On Photography*. New York: Farrer, Strauss and Giroux.

Sparks, Kira. 2013. "Digital Natives: Six Ways Marketers can Engage Millennials." 30 July 2013. Shoutlet. Accessed 16 September 2013. http://www.shoutlet.com/blog/2013/07/digital-natives-six-ways-marketers-can-engage-millennials.

Taylor, Mark. 1987. *Hiding*. Chicago: University of Chicago Press.

The Economist. 2010. "Technology and society: Is it really helpful to talk about a new generation of 'digital natives' who have grown up with the internet?" 4 March 2010. Accessed 22 April 2012. http://www.economist.com/node/15582279?story_id=15582279.

Urry, John and Jonas Larsen. 2011. *The Tourist Gaze 3.0.* Thousand Oaks, CA: Sage Publishers.

Veblen, Thorstein. 2009. *The Theory of the Leisure Class*. Edited by Matha Banta. New York: Oxford University Press.

Žižek, Slavoj. 2002. "The matrix: or, the two sides of perversion." In *The Matrix and Philosophy: Welcome to the Desert of the Real*. Edited by William Irwin. Chicago: Open Court, 240–266.

Žižek, Slavoj. 2009. *The Parallax View*. Cambridge, MA: The MIT Press.

Žižek, Slavoj. 2006. *Interrogating the Real.* New York: Continuum.

Part III

INTERFACES AND OTHER FACES

Chapter 8

Alterity and Technology

Implications of Heidegger's Phenomenology

Tales Tomaz

The question about alterity is the question about a relationship. How do we relate to the Other? What allows us to enter into this relationship? What makes, the Other, in fact, an-Other, not an extension of ourselves? Questions like these have occupied Western thought since the beginning of philosophy. However, the technological advances of the last 200 years and, in particular, the new technologies emerged since the mid-1970s that popularized so-called real time seem to require us rethinking all previous answers.

After all, with the technologies of real time, the Other attends in a way never imagined before. For the first time in human history, we can talk to people in real time, without their bodily presence. What kind of relationship is that? What do our technological ways of being in relationship with others tell us about alterity? And there is one that is perhaps the most disturbing question of all: who or what is in fact Other? What qualifies someone or something to be Other? With the emergence and improvement of artificial intelligence systems, we are taking firm steps toward a world in which we relate primarily with machines. In the relationship with them, can we also speak of alterity? What is alterity, after all?

This text intends to make a contribution to this debate by exploring implications of some phenomenological concepts exposed by Martin Heidegger. The theme can be discussed from different perspectives, and it is not the objective of this text to limit its possibilities to this one German philosopher. At the same time, we believe that Heidegger introduces very important concepts into the debate, even if he himself did not devote much attention to the theme of alterity. The text does not intend to address all the philosophical aspects of Heidegger, nor to follow him closely in all the details, including the absolutely inexcusable fact that Heidegger, guided by his own thought, once became involved with Nazism and anti-Semitism (Trawny 2014; Zimmerman 1990). Undoubtedly racist ideas found a foothold in his work and thought. So one must be careful in dealing with his thought, not merely reproducing it but trying to think with him in such a way as to think *against* him. Therefore, this text does not intend to echo Heidegger, but to focus on what seems to be essential in helping us today, in the twenty-first century, to address the issue of alterity. We hope that, throughout the text, it becomes clear why there is a need to consider Heidegger's phenomenology in order to think the relation between alterity and technology.

1. DASEIN: A BEING THAT RELATES TO OTHER BEINGS

We have access to the Other. The Other is present to us in some way. This means that we perceive it. Perceiving the Other seems to be a basic condition for entering into a relationship with it. This makes perception a privileged departure point to think alterity. But what exactly is perception? What does perceiving mean? What does it imply? Phenomenology took more seriously this question than any other field of knowledge. In his lecture *Basic Problems of Phenomenology*, Heidegger devotes a considerable part of his reflection to the problem of perception and begins by defining it as follows: "What we concisely call perception is, more explicitly formulated, the perceptual *directing of oneself toward* what is perceived, in such a way indeed that the perceived is itself always understood as perceived in its perceivedness" (Heidegger 1975, 79). Perception is therefore the *perceptual directing of oneself toward what is perceived.* This directing-toward-something is a typical structure that phenomenology describes as *intentionality*, common to all modes of comportments, such as representing, thinking, loving. Thus not only perception has an intentional character.

In scrutinizing the "directedness-toward-something," one can see that intentionality comprises at once two different moments, which are called *intentio* and *intentum. Intentio* refers to the directing-toward, while *intentum* is the toward-which we direct ourselves. In the case of perception, *intentio* is the perceiving, and *intentum* is the perceived

(Heidegger 1975, 83). What does this mean for our reflection? All intentional comportments always have some content, that is, they are already a mode of relating to someone or something. There is no pure representing or pure thinking. There is no pure loving. There must be someone or something to love. The same happens with perception. There is no pure perception. Perception is always perception *of something* and, therefore, it is a mode of relating to someone or something.

In the same work, Heidegger demonstrated the implications that this intentional character of perception has for an adequate understanding of the basic constitution of the human being. But to comprehend this point correctly, it is necessary first and foremost to avoid some misunderstandings that might occur with any reflection on intentionality. The first misunderstanding Heidegger (1975, 83–86) calls the *objectivist understanding* of intentionality. From this perspective, I assume that the object to be perceived and I exist both in the exterior world. Between this object and me appears a relationship—intentionality—which is what allows me to perceive it. This relation is established only when the two poles of the relationship are present. If I remove one of them—the object, for example—the relationship ceases to exist. Intentionality in this case could only exist if indeed there is a physical object before me. So we would have to conclude, in itself, an isolated subject has no intentionality. This is the objectivist way of understanding perception. Perception exists only when there is intentionality, and this in turn is an external thing that appears only when there are both of the poles of the relationship, the psychical subject and the physical object.

Nevertheless, this way of understanding perception and intentionality overlooks the fact that the subject is already in itself intentionally structured (Heidegger 1975, 84). The objectivist understanding of perception overlooks the *intentio*—the perceiving itself—that already belongs to intentionality. Intentionality is not something external, isolated, a third element that surges between subject and object. It has already the perceiving in itself, which means that the subject already perceives by itself. The subject has intentionality without the need for a physical object. Intentionality must therefore be localized in the subject, so to say. In order to confirm this argument, we just need to consider hallucinations. The person having the hallucination believes to see and hear things, although there are no physical objects around her. Intentionality, therefore, is something that is already present in the person. Intentionality is not an objective entity existing in the world outside.

However, Heidegger (1975, 86) warns that this can easily lead us to commit another mistake regarding intentionality and perception, namely the risk of thinking intentionality as property of an "I," of an immanent sphere of the subject, which needs a specific moment of transcendence to reach the beings outside. From such a misunderstanding can arise the question, "How does the subject relate to things outside?" How can it transcend its own sphere of subjectivity and establish a relationship with the beings-in-the-world out there? To avoid this trap, one needs to pay attention to the *intentum* belonging to intentionality. Perception is not directed toward mere subjective sensations (Heidegger 1975, 88). Perception is directed toward a perceived, toward the beings themselves. There is no subject in the sense of an "I" apart from the "world." We are, since the beginning, *in* the world. It is wrong to ask how things within the subject relate to the outside. This question presupposes a separation; it presupposes the subject as an "I" encapsulated in its sphere that somehow must transcend toward the world. This way of understanding the human being as an entity isolated from the world that must transcend to enter into relationship with things is derived from the Cartesian form of thinking, *cogito ergo sum*. But this interpretation ignores the originary character of perception: perception dwells already with the things. It always and already has been perceived. It does not need to provoke any transcendence. Things are already there where there is intentionality. To avoid any association with this Cartesian conception, Heidegger adopts the term *Dasein* when referring to the perceptional being, instead of subject. Although Dasein is a common word in everyday German, Heidegger ascribe to it another meaning by emphasizing its literal sense: *Being* (Sein) *there* (da). The human being is a Being-there, a Being-in-the-world. The implications of this modification will become clearer in the course of the arguments that follow.

Now, according to Heidegger, we are safe from misunderstandings with regard to intentionality. Intentionality is a mode of relating to beings. Better said, *intentionality is the very relationship with beings*. Therefore, any form of interpellation of the Other must necessarily pass through intentionality. Talking that way does not mean that the Other exists only as perceived, that is, in its perceivedness. It just means that the perceivedness of the Other is a necessary condition for access to and relationship with it. As a necessary condition, the question for Heidegger then turns out to be where to locate this quality of perceivedness of a being? The reasoning is similar to that of intentionality. If we think of any being, we will find many qualities. As an example, we can think of a computer. (Heidegger uses the window as example, but in order to approach gradually the subject of this essay, we can use the iconic example of our technological world). A personal computer has many features. It can be compact, fast, white, well-finished, etc. But the perceivedness of the computer is never found among them. Rather we say: the computer *is* the perceived itself. Hence perceivedness is not in the object, it is not something objective. But neither is it something subjective. After all, perceiving is already a directedness-toward the things themselves. The point of Heidegger's phenomenological argument is this: Perceivedness is neither in the object nor in the subject, but in Dasein's intentionality. Intentionality is what uncovers beings and allows them to be encountered. In Heidegger's (1975, 98) words, "perceiving is a *release* of extant things which *lets them be encountered.* Transcending is an uncovering."

That means that perceivedness is a mode of uncoveredness of beings *as beings*. But this uncoveredness of the beings must be found in the perceiving itself, otherwise perceiving would not be able to release beings *as beings*, for example, to understand the extantness of an extant. Perceiving must understand previously the extantness of something, in order to uncover it as extant. There must be a prior "understanding of extantness" in perceiving itself (Heidegger 1975, 99). To the intentionality of perception belongs therefore not only *intentio* and *intentum*, but also a prior "understanding of Being." The extant can only be uncovered as an extant, because Dasein already has a certain ability of disclosing the extant, that is, a previous understanding of what means to be.

This brief reflection on perception helps us see some important aspects that should not go unnoticed in any consideration about alterity and technology. It shows us that we, as beings who perceive, are always, from the beginning, in relationship with other beings. We do not exist as an isolated "I," which somehow needs to enter into relationship with the world and encounter the other beings. We are already *in* the world. That is why Heidegger says that "Being-there" (Dasein) is at the same time "Being-in-the-world." "World" here is not a reference to the planet Earth, but to the totality of meanings and references in which we are involved in our experience with things from the very beginning (Heidegger 1977b, 86). We cannot change the fact that, from the beginning, things already mean something. In this sense, we are thrown into a world which we have not chosen. But we can interfere with these meanings, change them, offering new significance. Aware of this possibility, we are always ahead of ourselves. This means that we are not tied in our present. Dasein is a permanent openness to its capacity-to-be. This condition ensures the understanding of Being and allows Dasein to encounter other beings and to enter into relationship with them.

Comprehending perception as an understanding of Being sheds light also on a very important theme for Heidegger that is also relevant to any thinking of alterity, namely ontological difference. With "ontological difference" Heidegger (1975, 322) means: the difference between Being (with uppercase) and beings. Exploring all aspects of ontological difference would require much more than a few lines. Here we focus only on what is essential. The point is that, as Being-in-the- world, I encounter beings *as beings*. Their "beingness" needs to be understood beforehand so that beings can appear as such. This beingness that reveals beings as beings is Being (with uppercase). One should not confuse Being with a supreme being, like God. Being cannot be an entity. Being is the revealing of beings as such. Without this difference between Being and being, it would not be possible to encounter beings as such, and consequently it would not be possible to enter into a relationship with them. This means that we can only enter into a relationship with things and with the Other because of ontological difference, that is, because we have a prior understanding of Being, or of what means to be.

This reflection on perception helps us to understand furthermore that there is a kind of entity that throughout history is in relation to the Being of beings, that is, a being to whom things appear as existents. Dasein is the human being. Not that the human being is Dasein, as if Dasein were a human property and the human were the foundation of Being. Instead, Dasein, the openness that allows the understanding of Being, happens in humans. Since Descartes, the idea has been that the human being is the foundation of Being. Understanding the Being—that is, understanding that things are and assigning meaning to them—is something that has been closely linked with human cognitive ability. Consequently, if we were able to reproduce this cognitive ability in some other kind of being, it would also be endowed with an understanding of Being. And that is, in fact, what is advocated by much of the theory of Artificial Intelligence (Leidlmair 1991). In this latter deployment, the Cartesian-modern comprehension seems to overcome anthropocentrism in our understanding of the real, appearing to situate all beings in a similar condition. However, it interprets Being as something that arises from the human—from human cognitive ability. Understood as Dasein, however, the human is not the fundament of Being, but the locus where the Being has occurred throughout history. This is a crucial difference (Trawny 2003, 52). This is a condition that the human being itself cannot overcome. It is not allowed to ask "why does it happen here?" Any question about what enables this understanding presupposes the understanding itself and is therefore a circular question. Dasein, the being where Being happens, is the human being.

2. THE OTHER AND THE "MAN"

The preceding has intended to demonstrate how the experience of a relationship with beings cannot be understood as a phenomenon that an isolated "I" performs, but as something that is already in the human being itself, due to its previous understanding of Being, that is, that things are. At this point, then, we are nearing an understanding of alterity. For this, it will be essential to better understand the "nature" of the beings that Dasein encounters in the world. As we have already pointed out, Dasein, thrown into the world, is already in relationship with beings. This makes it necessary to presume not only beings, but also the *understanding* of beings as such, that is, in their "beingness This understanding is already in Dasein, in the Being-in-the-world itself. The fact of encountering things as extants, for example, presumes that Dasein already possesses a prior understanding of extantness, which enables the uncovering of the extant as extant.

Something similar occurs regarding the Other. In its daily concerns, Dasein also encounters Others, who are always recognized as such. In the totality of meanings, the Other does not appear as another existing thing. The Other does not appear as extant, otherwise it would not be identified as Other, but as extant. "The Others who are thus 'encountered' in a ready-to-hand, environmental context of equipment, are not somehow added on in thought to some Thing which is proximally just present-at-hand; such 'Things' are encountered from out of the world in which they are ready-to-hand for Others—a world which is always mine too in advance" (Heidegger 1977b, 118). I actually encounter the Other *as Other in the world.* The Other is another Dasein. I see it as someone for whom things are also things, that is, as someone who is also, since the beginning, in a relationship with beings. When we see someone typing on a computer, however absorbed that person is in the device, it does not appear to us as a thing placed ahead of a second thing. It appears as Other.

This means that in the Being-in-the-world there is also the understanding of the "otherness" of the Other. This understanding uncovers the Other as another Dasein, who is (sein) together (mit) in the world (da, Welt). Thus, Heidegger (1977b, 117–118) calls the Other: Mit-Dasein, or Being-there-with. For this reason, we have to conclude that belonging to the understanding of Being is an *understanding of alterity*. One can see, therefore, that alterity is not something posterior in the experience of the human being, not even something that can be ignored. That the concept of alterity was underdetermined in Heidegger's work does not erase the fact that the original exposure of Dasein's condition already included the Other (Michalski 1999). Further on we will explore the need to deepen the concept of alterity in Dasein. What is necessary to emphasize right now is that alterity cannot be an arbitrary being; it cannot be any perceived. We encounter alterity in the other Dasein, in the being which is in the world just as we are. If Dasein is the human being, the logical conclusion must be that we encounter alterity only in another human being.

To move forward with the considerations on alterity and the relationship with the Other, we must also take into account another concept used by Heidegger, which will serve too as transition to move from alterity to the issue of technology. It is the concept of "Man," to which Heidegger devoted much attention in *Being and Time* but that seems to have disappeared in the later works. Heidegger affirms that, in its daily life, Dasein is assumed by the Man. *Man* is a German neutral word to indicate an indeterminate subject of a sentence. *Man sagt* means, for instance, *one says* or *they say*, when it is not certain who exactly says it. That means that "the others," "they" determine what Dasein is. For Heidegger (1977b, 126), this is the usual mode in which the Other encounters Dasein, that is, as "others" who determine its very Being. As Heidegger describes it, by using public transportation, by assuming our roles in the workplace and at school, by entertaining us in front of the television, each one is like the others. We take the bus and the subway as others do, we eat at restaurants as others do, we are outraged at corruption as others are. Thus, the others take away the Being of Dasein. Dasein is the others, so to speak (Heidegger 1977b, 127).

This means that it is not the individual itself who is "there," assuming its own responsibility to be—whether at work, home, free time etc. Its possibilities of Being are determined and restricted by others. Therefore, Dasein in this mode of Being tries to stay in the averageness, do not do anything eccentric, do not get away from what it considers reasonable. It constantly compares itself with others and tries to enter into relationship with others and with itself in a similar way. Attempting to keep itself in the averageness also leads to another fundamental point for Dasein: it levels down all the possibilities of Being (Heidegger 1977b, 127). By having its Being taken away by the others, by the Man, Dasein also experiences a tranquility [*Beruhigung*]. Dasein is tranquilized because it can cling to the certainties of the roles assigned to it and does not have to wonder about its possibilities of Being. Everything is "in the best of order" if Dasein follows the script that the others, the Man, ascribed to it (Heidegger 1977b, 177). Certainty and assurance provided by the Man lead to tranquility.

However, this does not mean that, in being tranquilized, Dasein will live in stagnation and inactivity. Because it needs to follow the script given by the others, Dasein is busy all the time. It has tasks to perform, and it must always ensure the performing of these tasks in the best possible way. The tranquilized state of Dasein is like a drug that always asks for more reasons to tranquilize. The more tranquility and assurance it wants to have, the more it will need to know and understand the world around it in minute detail. It will be moved to work with care and perfection, and this has a cost. It will need to work longer hours and more intensely, study its own functions, and update itself. Dasein will also be moved to better understand nature, and this has a cost. It will need to seek causes and effects in the natural world and, if possible, identify them also in the social world (Heidegger 1977b, 178).

Heidegger calls this whole mode of Being "decadence" [*Verfallenheit*]. In the way he describes it, decadence seems to be a negative condition. The very use of the term "decadence" suggests a negativity. However, Heidegger (1977b, 175) affirms that this concept is not intended "to express any negative evaluation," cannot be understood "as 'fall' from a purer and higher 'primal status'" (1977b, 176), as the concept of "fall" in the Christian tradition suggests. Decadence is the condition of Dasein in its daily worldly concerns, because it plunges out from its whole capacity-to-be to a specific mode of Being in order to live in society. Leidlmair (1991, 165) indicates that we should understand the

"Man" in Heidegger as a "prescriptions of roles." We fall prey, he says, to these prescriptions of roles. When people are in the mode of Being of decadence, taken by Man, that is, in their daily concerns and social lives, they are not individuals, but see themselves as parts that perform a function. What they are is determined by the role they assume.

The roles people play even have an economic function within different social relations (Bammé et al. 1983, 164). On public transportation, people are users. In news consumption, they are viewers or readers. While in the production of news, they are journalists and reporters. When one assumes the role of journalist, for example, it is expected that s/he acts in a certain way, that s/he behaves in accordance with the assumed role. In presenting a news program, it is not expected that the journalist will say: "Today there was nothing relevant, so there will be no news. We will come back tomorrow, if something important happens." This is not part of her/his role. S/he is there exactly to present things that are supposed to have relevance and ensure that the news occurs daily. Likewise, in the operating room, a surgeon is not expected to begin to pray with us or to say: "Today I am not in the mood to work." It is expected that s/he begins to operate immediately. Thus, thanks to the mode of decadence, it is not necessary to spend hours in a dialogue between surgeon and patient so that everyone understands what to expect from each other. That is why the Man represents an economy in social life.

For Heidegger, this is how we enter into relationship with the Other in our daily concerns. Thinking alterity this way, however, leaves room for a very pertinent criticism. According to S. J. McGrath (2008, 80), Heidegger proposes a kind of "collectivism" as opposed to "liberal humanism." McGrath draws attention to the fact that, in Heidegger's understanding of Dasein, the Other is not something radically different. Luckner (2008, 87) offers similar criticism by saying that, for Heidegger, only death is something really distinct from Dasein and able to reveal its finitude, while the Other and other things are mere moments of its own existence. In fact, in *Being and Time* Heidegger (1977b, 118) states that the others are "those from whom, for the most part, one does *not* distinguish oneself—those among whom one is too." In Heidegger's comprehension, then, the Other would not be an irreducible difference. McGrath (2008, 81–82) adds, "for Heidegger, on the other hand, Dasein knows exactly what to expect from the others insofar as they are no different from it. The others do not confront Dasein; on the contrary, they allow Dasein to avoid the confrontation with itself that is always impending."

For Levinas (1999), the face of the Other is the face of someone I do not understand, which imposes limits on my own Being. It is the realization of the limit of one's own subjectivity that forces it to recognize the individuality of the Other, the value of intersubjectivity and the need to share and give in. In Heidegger there is a continuing risk of ignoring the intersubjective because of the necessity of differentiating oneself from others. Only this differentiation grants to Dasein its ownmost possibilities as the capacity-to-be. It is not surprising that this interpretation of Dasein easily legitimates the rule of the strongest—those who distinguished themselves—over the weak—those whose Being is determined by others—which puts Heidegger in line with an elitist tradition in Western thought. For him, taking care of the Other in order to reduce its afflictions removes its own occupations and makes it dependent and dominated. This is necessarily a bad thing *from the point of view of an ownmost existence*. The "ownmost" care would instead make the Other see in its problems its own existential condition of being thrown in the world, and be free for its ownmost possibilities (Heidegger 1977b, 122). The fact that Heidegger chooses this understanding of "personhood" is undoubtedly one of the doors that opens onto Nazism in his work.

At the same time, it is important not to ignore the theoretical gain that the concept of Dasein provides to think alterity as a factual experience of human existence. Thus, instead of restricting the encounter with the Other to Man, as Heidegger did in *Being and Time*, one must see the Other in its more original sense, that is, as the one who also understands Being, understands that I am and that it is. *I know that it knows*. Interestingly, it seems that Heidegger did not realize the appeal of the Other as Mit-Dasein: *I know that you know. We share the same world.* For some, this world can be more transparent than for others. For some, the "understanding of Being" as a precondition of their existence can be clearer than for others. But the Other—in ownmostness or not—is involved as well from the beginning in the totality of meanings and references of existence. It is also a capacity-to-be and is in itself a universe which imposes a limit to my existence. Recognition of this limit *is given in advance in Dasein*. This limit must be *suppressed* through a modification of our originary condition if we are to see the Other as something to be manipulated or if we want to deliver ourselves over to the Man. In short, if we are to reduce our relationship with the Other to the prescription of roles. In this sense, the Other as Mit-Dasein requires an expansion of the original meaning that Heidegger wanted to give to it in *Being and Time*. This expansion would avoid the risk of elitism always present in Heidegger's trajectory. Faced with the tragedies that the elitist thinking caused in the last century and facing the challenges that it presents in the current century for the world and especially for Europe, it is important to consider even more radically that the Other is a Mit-Dasein *who we always recognize* and who we can only escape by escaping from our own originary condition as the being that relates to Being.

3. TECHNOLOGY AS A MODE OF BEING

So far, we have seen that Dasein enters into a relationship with the Other because of its fundamental constitution as Being-in-the-world. Dasein is thrown into the world, dwelling from the beginning with things and Others. This means that it is able to differentiate Being and beings, that is, to understand beings *as beings*. Dasein is therefore the locus where Being uncovers beings. This uncovering happens in the human being, once the human has an understanding of Being. The Other is the other human being, who we encounter as such since the beginning in our factual existence. At the same time, Dasein has a tendency to let itself be determined by others, the Man. However, these others determining Dasein are not the Other understood in its originary condition as Mit-Dasein. In the Man, the Other is already reduced to the role it plays. And, as we shall see, this reduction of the Other is exacerbated in technology.

The question of technology appeared solely in modernity, with the machines brought to light by the Industrial Revolution. Since then, the issue has increasingly been the subject of attention in various fields of study and research. Media and communication studies are, of course, among the most interested in this discussion, after all, these new technologies have often found their most effective consolidation in human communication processes, although not entirely restricted to them. As already stated, we increasingly communicate with Others without the immediate presence of a physical body, creating an unprecedented experience in human history. All of this is centered on specific equipment, which therefore draw our attention. Such equipments (computers, laptops, smartphones, etc.) and functionalities (news portals, search engines, social networks, instant chat applications, etc.) are so-called "new media." For enthusiasts, the presence of these devices is vital to progress. For others, who issue a more apocalyptic warning, the relationship with such devices is considered destructive, eroding everything that we value. On one side or another, the focus is always on the devices themselves. *They* are called *technologies*. But what happens when such equipment is increasingly miniaturized and incorporated into our own bodies? What happens when there is a generation of people who will be a mixture of flesh and chip, proteins and bits? Can we continue talking about technologies? Will it still make sense to question the relationship between alterity and technology? Will it still make sense to speak of a *technological world*?

Dealing with these questions depends on the definition of technology. Heidegger (2000, 9) made an essential contribution to this debate by developing a reflection that shows that "the essence of technology is not technological." It is right to say that technology is a means or an instrument. But what is a *means*? It is a way to do something. As such, it is a mode of relating to the real. Employing a means expresses and reinforces at once what is meant by the real. Therefore, a means is always a form of constitution of world. Seubold (1986) illustrates this condition with the example of the Indians of Taos, in New Mexico. When the Spanish settlers introduced the plow to them, the equipment was immediately rejected because, for those people, it hurt the breast of the earth. By the same reasoning, this tribe removed the horseshoes from its horses during the spring so as not to hurt the pregnant earth. In both cases, the Indians of Taos employed different means to do seemingly the same thing that the Spaniards were doing, namely cultivating the land. But they were not simply employing other means; the employed means evidenced a different relationship to nature and the world, in short, to the real. A new means requires a new relationship to things or a new way of perceiving things from a different angle. For the Spaniards, the land was an assemblage of minerals at their disposal and available to them in order to increase productivity. For the Indians of Taos, the earth was a mysterious entity, autonomous, almost a deity, which in its generosity benefited human kind by offering its fruits.

This example shows that the fundamental aspect of technology as a means of doing something is not located in the equipment itself, but in the relationship with the real that it expresses. Once our entering-in-relation with things and Others depends on a previous understanding of

Being, namely, an understanding of what the entities are, then what is decisive in technology is the *understanding of Being* that it bears. It is a specific mode of Being and, therefore, a way of acting. The question should then be, "What is the understanding of Being that technology bears and lays bare?" Heidegger (2000, 20) explains that technology, as a mode of Being, uncovers beings as objects at our disposal, in their "standing reserve" or *disponibility* [Bestand], so that they can be manipulated in order to achieve certain results. In the example above, the technology of the plow unveils the earth as minerals that are available to enhance the production of food. Technology is therefore a mode of Being and acting in which things appear as objects that can be used to perform other activities. This technological action is possible only because we have a prior understanding of Being that enables us to enter into relationship with things. The understanding of Being is what allows us to encounter beings, whether as things, as Others, or as standing reserve or disponibility. As with *das Man*, beings are not encountered in their autonomy and independence, but in the role they play in performing a specific task. By giving things a specific purpose and regulating their use, technology saves time and effort. It is an economy. But at the same time it also restricts other possibilities of accessing things and the Other, because they turn out to be more laborious, risky and finally *unreasonable*. Technology restricts the modes of being and acting of everyday life for the sake of modes that run as economically and automatically as possible (Luckner 2008).

This character of technology and its similarity to *das Man* becomes clearer with the conception of the machine that has emerged since the appearance of cybernetics, informatics and the computer, which Bammé et al. (1983) call the "transclassical machine." In this new concept, the machine is understood as a behavior. A calculator, for example, is a device to perform calculations. There are several types of calculator: mechanical, electrical, electronic, etc. However, there is something in common between all of them, which is really what is decisive. They all follow a "step-by-step" process to transform the inputs into a specific result. This step-by-step process actually does not even need equipment to be performed. We all learn in school how to perform addition, for example. This step-by-step approach that the calculator (or ourselves) needs to follow in order to perform the calculation is what is called an *algorithm*. Algorithms are precise logical instructions that require no interpretation. It is a procedure that represents an economy of thought and effort. The physical device is only the material embodiment of an algorithm (or a set of algorithms). What is important, then, is not so much its materiality, which could be any number of things, but the procedure it performs, its behaviors (Bammé et al. 1983, 115–145). This became even more apparent with the computer, a device that, thanks to advances in formal logics, can run *any* algorithm, that is, can perform as many different machines as is possible (like the "universal machine" that Turing had theorized).

When we restrict ourselves to calculate things, for example, we are proceeding in a machine-like fashion. As Luckner (2008, 44–45) highlights, this machine-like procedure is only possible because we have the ability to relate to things and comport ourselves in different ways, including with specific formal procedures that can be followed without requiring thinking or interpreting. The relationship between surgeon and patient, remembering the example cited above, depends largely on following a specific formal procedure that, in the light of the foregoing, may be described as mechanical or *machine-like*. But we only have the ability to see beings as objects and use them in machine-like ways because we previously have had the understanding of Being that gives us several possibilities of comportment, *including a machine-like one*. The machine itself is a *mode of Being derived from the originary condition of Dasein as openness. Das Man*, technology, and machine all correspond to the same thing—a mode of entering into relationship with things and with the Other that restricts the originary mode of Being of Dasein to specific forms in order to facilitate daily life (Leidlmair 1991).

What is the matter, for Heidegger, is that we have reached a point in human history where technology prevails in such a way—by generating economy and efficiency through formal procedures—that it becomes the only legitimate modes of Being, restricting all other modes of being and all other ways of relating to things. Things, therefore, tend to appear only as standing reserve or disponibility. Everything is reduced to the role it can play. In modern societies, that is, in the age of technology, we have a tendency to reduce all our actions to formal behaviors, to machine-like procedures. This tendency becomes a self-fulfilling prophecy—reduced our Being (capacity-to-be) to machine-like aspects, we cannot distinguish ourselves from technological equipment *that appears to behaves like we behave*. Hence the tendency to design "friendliness" (i.e., "a user friendly interface") into our technological equipment develops not because of the technological advance in itself, but because of its progressive success in reducing our comprehension of the complexity of our existence and social life. As an example, it is worth mentioning the well-known case of the computer program "ELIZA." Created in 1966 by Joseph Weizenbaum, an engineer at MIT, the program sought to simulate a dialogue between a patient and psychologist. The program employed the user's own words to formulate responses so that the conversation could proceed without the computer needing to understand anything or acquire any information. The user could write, for example: "I'm having trouble with my father." In response, the program could reply: "Tell me more about your family" or "With whom else do you have difficulties?" It had this ability by identifying certain keywords in the user's sentence and then replying with ready-made answers associated with those words, usually based on a dictionary (where "father" appeared, for example, as a member of the "family"). The procedure was extremely simple. Weizenbaum himself said that his program was a "parody" of the relationship between patient and psychologist. But it was also a parody of human conversations in general. The choice of patient-psychologist relationship occurred only because this relationship is socially allowed to show no real knowledge or interest in the world of the Other. Consequently it did not matter if the answer came from another human person or a computer. To Weizenbaum's surprise, however, the program was welcomed by many as a real breakthrough, being taken seriously by many users (like his own secretary, who reportedly talked with ELIZA for hours), who thought that the program could really understand their problems. The situation eventually came to be called the "Eliza Effect" (Hofstadter 1995, 157).

It is true that the web, thanks to the huge amount of data we are feeding search engines and social networking sites, is making these algorithms more and more capable so that they present behaviors that appear to simulate human intelligence (Shroff 2014). But when we expand the understanding of what a machine is, defining it as the algorithm itself, it is clear that its success lies not so much in its increasingly complexity and sophistication, but in the progressive reduction of modern human life into procedures or algorithms. This is also true for contemporary ELIZAs, like Siri and S Voice. They correspond more and more to our needs certainly due to improvements in the design of the algorithm, but also because of the fact that our life has been simplified and codified into logical processes and formal structures. Even the interpellation of the Other in new media works because we have already reduced what the Other is. All its physical and existential complexity is reduced to expression in signs. For this reason, we are satisfied to have access to the Other in video, audio, or even text messages. As an assemblage of signs, the Other is something available at our disposal; it is a resource to which I resort, when I am in need. In this sense, it matters little whether there is or is not a physical, bodily Other behind the signs. The important thing is the *function of alterity that it performs*. In addition, the web and especially social networks, while providing us with vast and transparent reservoirs of the practices of others, also offer an unprecedented and growing amount of automated modes of how to be and act in specific situations, for example, which places to travel to, which photos to take, what to eat, which authors to read, and so on. The list is virtually infinite. Although increasingly diversified, these are modes of being with and relating to the Other that restrict possibilities. And this is only possible because of the reduction of the Other and of ourselves to the fulfillment of social roles. Therefore, the use we make of the Other, transformed into assemblage of signs in new media, is *essentially* technological and indicates the reduction of the human as capacity-to-be.

4. THE TECHNOLOGICAL PRODUCIBILITY OF EVERY BEING

It is clear, therefore, that what is decisive in technology is that it is a *mode of Being* of Dasein, a mode of coming into relation with things and with the Other. But for the question concerning alterity, what is most relevant is the ontological origin of this mode of Being. That is, where does the tendency to develop this mode of Being come from and what does it mean regarding our originary constitution? It is only in addressing this question that we can begin to correctly interpret alterity in technology. We can, for instance, operate in the machine-like mode of Being and acting, because we already have the ability to turn everything into an object at our disposal. According to Heidegger, this ability derives from our tendency to interpret Being from the beings that we encounter in the world, and not from our originary condition as an openness that understands Being. This tendency stems from Being itself. By revealing beings, Being hides itself, in much the same way that the surface of a mirror cannot be seen in order for it to work *as a mirror*. "The unhiddenness of being, the clarity granted to it, darkens the light of Being. Being evades by unconcealing itself in being" (Heidegger 1977a, 310) Although the ontological difference—between Being and being—is precisely what allows us to encounter beings and enter into a relationship with them, this difference is veiled at the time that we encounter beings. There seems to be only beings—no Being. Then we tend to interpret all beings as equal, including ourselves, and we forget that it is the event of Being in us that allows us to encounter them. We tend to forget the ontological difference, Being and our originary condition as beings in which Being reveals beings.

This forgetfulness reaches a crucial point when Leibniz formulates the principle of reason: *nihil sine ratione*, nothing is without reason. Insofar as there seems to be only beings, everything seems explicable as a causal chain of relations between beings. Then comes the realization that nothing is without reason, nothing is without cause. By formulating this principle, Leibniz condenses more than two thousand years of the history of Western metaphysics, namely the search for reason, the cause, or the foundation of things. Why does metaphysics want to know reason and cause? To ascertain things as objects, that is, to be able to count on them (Heidegger 1997, 50). In order to be able to count on an object, we need to know it in every possible detail, everything that makes it up, that can influence its position or movement, its current state or change. In the knowledge of its causes, namely its conditions of possibility, an object can be assured. If one can determine everything that makes it possible—everything that allows it to be—its condition will be assured.

Everything has a reason; every being has its conditions of possibility. Nothing is without cause. The most important thing to realize in this principle of reason is that the cause, the conditions of possibility, is converted into the very meaning of the word *Being*. Beings are what have a cause. If this cause is unveiled or revealed, it becomes possible to count on the being and even to (re)produce it. Hence the need to research endlessly in search of the cause of any phenomenon. In this movement, however, there is a confusion between Being and beings. The causal chain necessarily leads to the search for the ultimate cause, and this also appears as an entity, a being. We forget that the beingness of things is only known because it is already given from the beginning. Being is something that we cannot overtake. Western metaphysics operates by interpreting Being as the condition of possibility and thus enables the forgetfulness of the difference between Being and being and of Being itself.

The interpretation of Being as a condition of possibility is precisely that which lies at the base of technology. "The conditions of possibility of technological producibility are simultaneously the conditions of possibility of the Being of beings" (Leidlmair, 153). In the age of technology, being is what can be produced and Being correspond to the conditions of producing; being is that which can be explained step-by-step, assured, verified, and therefore subject to (re)production, inasmuch as one discovers its conditions of possibility, its causes. However, in the age of technology we forget not only Being, but the very forgetfulness of Being. We seem to have reached the end of history. That is, everything seems to have finally been discovered. Nothing seems to be hidden, and it is not expected any significant reversal. The mathematical-technological topography is meant to be the universal itself (Trawny 2015). There is nothing beyond technology. Before we question whether it will be possible to speak of the technological world when we incorporate in ourselves chips and bits, the answer is already available: *The technological world is the world itself.* We do not come into or leave this world by adhering to or refusing this or that equipment. We do not give in to this mode of Being of our own will. Things are revealed as mere disponibility (or "standing reserve," which is the more standard English translation of Heidegger's *Bestand*) in that they respond *in this way* to our dealings with them. Technology works. Likewise, we will not stop seeing things technologically of our own will. We will not interfere with it by a collective decision to resist becoming cyborg. Such an Event depends on the possibility that things show themselves otherwise. However, none of this erases the fact that the technological world, our world, is founded on a forgetfulness of the originary concealment of Being. We forget that every mode of the presentation of beings is already a concealment of Being, in whose light beings are unconcealed.

5. CONCLUSION

Technology in its current stage is a mode of Being in which all beings are interpreted as extant beings, *including Dasein.* Strictly speaking, there is no Dasein. The fact that humans have a relationship with their being and so enter into a relationship with the Other is merely a circumstantial issue that can be comprehended, manipulated, and therefore reproduced in another being. Dasein comes to be in the hands of humanity, which turns out to be seen as the foundation of Being. Thus, the human being forgets its originary condition. It believes itself to be as producible as any other being, as long as it finds its conditions of possibility. It also believes that it is capable of creating a being that, *in essence*, is the same as the human being. And indeed, operating within this paradigm, it is increasingly possible that this prediction will be fulfilled, because the *essence* of the human happens to be determined by technology. This uniformity between the human and other beings becomes the seal of the complete mastery of technology (Heidegger 1977a, 102–103).

In this way, therefore, human beings are seen as mera extant beings, not as Daseins, and alterity is perceived to be a relationship that arises between them. I look at my relationship with the Others from the point of view of a third party. Then, why not assign alterity also to other beings that increasingly behave in similar ways? There is a problem indeed. Thinking alterity as something that can exist between two extant things is already thinking it within a technological frame, within the technological paradigm. It is an epiphenomenon of technology and of the forgetfulness of the originary concealment of Being. Besides that, it is not due to human will that one can decide who or what is alterity. This is a determination of Being, insofar as Being originary discloses the Other in the other Dasein. Trying to see alterity in other being is misunderstanding our peculiar relationship with Being. But the paradox is that *Being itself* is dragging us onto this path in that the human dissolves in the formal and logical procedures of technology. It also becomes standing reserve or the disponibility of technological progress, as Heidegger (2000, 30) foresaw. In this way, Self and Other can be any and every being in the chain of reproducible causalities. It is indeed an explosion of selfness and otherness in all directions. But we should not forget what explosion at last means for the thing itself. Its final point.

REFERENCES

Bammé, Arno, Günter Feuerstein, Renate Genth, Eggert Holling, Renate Kahle, and Peter Kempin. 1983. *Maschinen-Mensch, Mensch-Maschinen: Grundrisse einer sozialen Beziehung*. Hamburg: Rowohlt.

Heidegger, Martin. 1975. *Die Grundprobleme der Phänomenologie*. Frankfurt: Vittorio Klostermann.

Heidegger, Martin. 1977. *Holzwege*. Frankfurt: Vittorio Klostermann.

Heidegger, Martin. 1977a. *Sein und Zeit*. Frankfurt: Vittorio Klostermann.

Heidegger, Martin. 1997b. *Der Satz vom Grund*. Frankfurt: Vittorio Klostermann.

Heidegger, Martin. 2000. *Vorträge und Aufsätze*. Frankfurt: Vittorio Klostermann.

Hofstadter, Douglas. 1995. *Fluid Concepts and Creative Analysis: Computer Models of the Fundamental Mechanisms of Thought*. New York: Basic Books.

Leidlmair, Karl. 1991. *Künstliche Intelligenz und Heidegger: Über den Zwiespalt von Natur und Geist*. Munich: Fink.

Levinas, Emmanuel. 1999. *Alterity and Transcendence*. London: Athlone Press.

Luckner, Andreas. 2008. *Heidegger und das Denken der Technik*. Bielefeld: Transcript.

McGrath, S. J. 2008. *Heidegger: A (Very) Critical Introduction*. Michigan/Cambridge: Grand Rapids.

Michalski, Mark. 1997. *Fremdwahrnehmung und Mitsein: Zur Grundlegung der Sozialphilosophie im Denken Max Schelers und Martin Heideggers*. Bonn: Bouvier Verlag.

Seubold, Günter. 1986. *Heideggers Analyse der neuzeitlichen Technik*. Freiburg/Munich: Alber.

Shroff, Gautam. 2014. *The Intelligent Web: Search, Smart Algorithms, and Big Data*. Oxford: Oxford University Press.

Trawny, Peter. 2003. *Martin Heidegger*. Frankfurt/New York: Campus Verlag.

Trawny, Peter. 2014. *Heidegger und der Mythos der jüdischen Weltverschwörung*. Frankfurt: Vittorio Klostermann.

Trawny, Peter. 2015. *Technik.Medium.Kapital: Das Universale und die Freiheit*. Berlin: Matthes & Seitz.

Zimmerman, Michael. 1990. *Heidegger's confrontation with modernity*. Indianapolis: Indiana University Press.

Chapter 9

Alterity ex Machina

The Encounter with Technology as an Epistemological-Ethical Drama

Mark Coeckelbergh

Alterity is a term used by philosophers to refer to "otherness" in relations and encounters between humans, most notably in the philosophy of Emmanuel Levinas (1961; 1995). For Levinas, humans have a radical difference. They are neither objects nor mere means; they have "face" and call for a response. But could the term also refer to (aspects of?) relations and encounters with non-human others such as animals and even machines? Is there alterity in the encounter with technology, and in what sense? Interestingly, Don Ihde (1990, 97) already asked in what phenomenological senses a technology can be other, and argued that our relation to technology can take the form of alterity, where the technology appears as a "quasi-other" (98). For instance, a robot may appear as a quasi-other – and indeed quasi-human. For Ihde, alterity is a particular kind of relation between the "I" and "technology," in which the technology emerges as a focal entity, a "focal quasi-other with which I momentarily engage" (107). But is this the only way we can understand technological (quasi-)alterity, or is there more to be said and told? A further discussion of technological alterity, including machine (quasi-)alterity, may contribute to (post)phenomenological investigations of technology. This chapter focuses on the machine and explores further meanings of machine alterity by putting the issue in the context of a wider field of questions concerning our relation to the other and non-humans, including (1) the exercise of defining the human by means of technology and other others, and (2) the phenomenology, epistemology, and ethics of alterity, with a particular emphasis on problems concerning the appearance of, and our response to, the alien.

First, the chapter shows that in Western philosophical anthropologies and ontologies, all kinds of others have been used to define the human. In modern times, for instance and perhaps most notably in the philosophical anthropology of René Descartes (1637), the machine was used as the other of the human. It is argued that this negative anthropology (Coeckelbergh 2009; 2014) always excludes other meanings of both the machine and the human. In the operation that constructs and uses them as the other of the human, the humans and non-humans involved are reduced to sameness, to general categories such as "animals" and "machines." I explore how we can respond to this problem and suggest that a Levinasian turn to the concrete encounter with the other may help.

Such a turn, however, raises its own problems. Therefore, secondly, the alterity dimension of the encounter with technology is further discussed by asking the question how the other can appear at all, especially if otherness *as alienness* is in question. In particular, it is argued that the encounter with machines may indeed assume the aspect of otherness and alien(ation), but that it is difficult to preserve or even experience this otherness and alienness, since usually machines move from the categories of "alien," "uncanny," and "monstrous" to "familiar," "domesticated," "normal," etc. – in other words from other to same. This may exclude *other* ways in which the machine can appear to us, other meanings of the machine that have the aspect of alterity as alienness. It is then asked more generally how the truly alien and other can appear to us at all, given that we immediately try to categorize it and domesticate it – for instance, in the case of the machine by our use of language and by our use of the technology. Can we avoid threatening alterity, or do we always domesticate and humanize? Does hospitality (Derrida 2000) toward the other erode the otherness, or can some place for otherness be saved?

The answer given in this chapter is that we should not assume full control of the phenomenology and epistemology involved in the encounter with any entity – here the machine; some meanings and (quasi-?)alterity faces of the machine may visit us uncalled, break in. In the drama of appearance and disappearance, which is so typical of the phenomenology and epistemology involved here, the machine may surprise us. Perhaps we have to await this *alterity ex machina*, rather than trying to control it. And could it be that true hospitality and ethical response *presupposes* (leaving room for) such moments of surprise and alienation, such moments in the encounter that escape language and *episteme*, such moments when alterity shows itself – here in and through the machine which may suddenly escape its identity as a machine and reveal itself as (quasi-)otherness? In the course of the discussion references are made to work by Ihde, Levinas, Derrida, and Heidegger, as well as to the tradition of romanticism, to suggest an answer to the question asked in the chapter. Although the focus is on the machine, the present essay also aims to make a contribution to more general discussions in the areas of phenomenology and ethics of alterity and alienness, and indeed to currents of thinking that connect phenomenology and ethics.

NEGATIVE ANTHROPOLOGY: THE MACHINE AS THE OTHER OF THE HUMAN

Before looking into the question concerning alterity in concrete encounters between humans and machines, it is helpful to first consider exercises that *abstract* from such concrete encounters and make general arguments about what humans, animals, and machines *are*: Western philosophical anthropologies and ontologies. As I have argued previously (Coeckelbergh 2009; 2014), Western anthropology tends to defend what the human is by using a *via negativa*, by saying what the human is not. In other words, it is a negative anthropology (compare to negative theology). In particular, what the human is, is constructed by using various non-human others: various animals, machines, and other non-humans (fictional or not) are constructed and used as the "other" of the human. Humans are defined as non-gods, non-angels, and non-animals. Other others are, for example, wolves, werewolves, dragons, monsters, and vampires. In each case, some sameness is presupposed – there are similarities with the human, otherwise the exercise would not work – but alterity is emphasized since the non-humans are seen as the *alter* of the human, and the non-human others are seen as the negative of the human.

In the context of discussions about robots, machines, artificial intelligences, automata, and so on, non-human alters include for instance the Golem: a human-like being that is created from inanimate matter and is described in for instance a famous narrative that involves a rabbi of Prague who made the Golem out of clay and animated it. And of course various automata and machines are used as non-human alters. Descartes already used machines as the negative or other of the human. In his *Discourse on Method* (1637), machines do not appear accidentally, they are a key part of his anthropology and philosophy of mind. For instance, Descartes argued that a machine can "utter words" but that "it never happens that it arranges its speech in various ways, in order to respond appropriately to everything that may be said in its presence" (Descartes 1637, 38). The human, then, appears as the negative of this machine: humans can arrange their speech in various ways in order to respond appropriately to what is being said. Humans respond in a non-machine-like way. The machine appears as the negative of the human and the human appears as the negative of the machine.

Today the computer, aliens, zombies, and robots are used as others to define who/what we are. Again the other can be imaginary or not. For instance, there are robots in literature, film, and philosophical thought experiments, but there are also *material*-anthropological exercises that take the form of designing and building the quasi-human. The design and engineering of human-like robots can be interpreted as a practice that helps us to understand the human. It seems that we need and use various "alters" to define ourselves. At least in the West, human-like creatures are engineered to imitate the human (and models of the human are used that are based on the machine), but at the same time the actual machines clearly show and emphasize the differences between humans and machines. Using all kinds of entities (animals, machines, etc.) we construct the human.

An ethical problem that may arise here is that we indeed *use* these entities for our anthropological and ontological exercises, that we use them to say something about the human but disregard their own potential otherness that may escape that relation. We impose our definitions on them, and this may do violence to them in the sense that there is no recognition of their alterity, or rather, that *if* they might have an alterity aspect, it cannot appear and is suppressed. They are "others" in relation to us, in the sense that they are "negatives" of us and *alters*, but do not get the chance to be taken seriously in their own otherness, which potentially they might have. At least we cannot *know* if they have it, if we treat them as negatives of the human. Moreover, differences between these "non-humans" are wiped out. All animals, all machines, all monsters, etc. become all the same: non-humans. To avoid this *Gleichschaltung* by anthropological and ontological means, and achieve a more ethical relation to these non-humans – that is, to render it possible that they cannot appear as other than non-humans – it seems that we need to turn to the concrete encounter with the other.

With Levinas (1961; 1995) we can turn to the encounter with the other, but against Levinas we can explore alterity aspects of the non-human other, which is always potentially more than a "non-human." Levinas emphasizes the transcendence of the human, but at the expense of the "non-humans" that are used to construct this transcendence and that are silenced. We can do better. For instance, Coeckelbergh and Gunkel (2014) have argued that animals can have "face." There are processes of facing animals and defacing animals, and these have ethical consequences.

In these processes language plays an important role. For example, for our ethical relation it matters whether or not we name animals. An animal with a name is usually treated better than an animal without a name. And as Jacques Derrida has suggested, the very word "animal" is ethically problematic. (Derrida 2008) In my book *Growing Moral Relations* I already argued that language is one condition of possibility of moral status ascription: our thinking about moral status depends on language and on the relation we have with the entity (Coeckelbergh 2012). Our moral relation to animals depends on the relations we have with them, including the concrete encounters with them and how we perceive and experience them in these encounters.[1] Similarly, one could argue, our moral relations to machines depends on their appearance in the concrete encounters we have with them. Here too language plays a role: if we already construct them as "machines," it seems that given our modern anthropologies they can only appear as such, as a machine and as the other of the human. But there *might* be a different "otherness" to them, which escapes our anthropological and ontological exercises and categories, and which might show itself in the concrete encounter with the machine – or more-than-machine/other-than-machine. At least it seems that, in order to avoid the violence diagnosed previously, we need to have an ethical starting position that at first leaves open the question of machine otherness, that does not start by asking or answering the "is" question – for instance that does not close down the discussion by arguing that machines *are mere machines* and that therefore *they cannot have otherness*; it is through this kind of operation that so much violence has been done to humans and animals, and it seems at least recommendable and desirable that we move on to a different, *other* kind of thinking. Let me further explain and discuss this in the next section.

THE ENCOUNTER WITH THE MACHINE AND THE (DIS)APPEARANCE OF OTHERNESS

Let us start with a more general question concerning alterity. At first sight it seems entirely plausible to say that otherness may appear in concrete encounters. On second thought, however, this is rather mysterious: we have to ask the philosophical, epistemological question how alterity can appear at all in the encounter with the other, given that we tend to try to turn otherness into sameness. Before turning again to the machine, therefore, let me first explain and develop this thought by considering how the truly *alien* may appear at all.

Consider the issue of an alien, extraterrestrial intelligence. *If* there was such an entity, how could we recognize it? If it were truly alien, then it seems that it would be difficult to know and experience it, indeed encounter or confront[2] it, given that our epistemic capacities are tied to the familiar. The only way we could get to know it, then, is if we *familiarize* ourselves with it. Then we might be able to represent it, communicate with it, and so on. But if it becomes more familiar, it also loses its alien character. The encounter becomes less of an encounter and meeting. Once the entity enters our world (not in the sense of a reality-in-itself but rather a reality-already-perceived-and-shaped-by-us) and indeed once we invite it into our world, is no longer entirely alien and radically other. It becomes more like the non-humans mentioned in the previous section; indeed, it becomes non-human, that is, it comes to be defined in relation to us. And as social and linguistic relations develop, as it becomes categorized, represented, and socialized, it becomes questionable how we can still think and perceive its alien aspects, its non-representational sides, and indeed its alterity. Ethically speaking, it becomes questionable if we can do justice to its alterity, if we can still respect the alien as alien once it enters the *domus*, the home, and the family. It seems that, from the perspective of a Levinasian ethics of alterity at least, here there is *too much* familiarity and sameness.

I propose that the same process happens with technology, in particular with machines that may invite the thought that we can enter into what Ihde (1990) calls an "alterity" relation with them. When new technologies are developed, they may first appear (to users) as somewhat alien, wild, and even uncanny. They will have the aspect of alterity and difference. But then there is the process of *domestication* of new technology. As we familiarize ourselves with the technology, the alien becomes familiar, the wild becomes domesticated, and alterity largely disappears. The machine becomes a quasi-other in the sense of an entity we live with, (socially) relate to, but it is no longer an other in the sense of radically different, alien, unfamiliar, and so on. It becomes more of the "same." There is no longer an en-counter, a meeting. We are no longer challenged. We are no longer called. We have already named and categorized. We have already domesticated. This may be ethically problematic, in the sense that if we miss *other* aspects of the machine, we could at least potentially do violence to it by reducing it to whatever name we give it. Through our language and social relations, we take the machine into our home. But by doing this, it may also lose aspects that first made it appear alien and other. Something is lost from our experience, and something is done to whatever the machine was before it entered our home.

In other words, from the perspective of an ethics of difference and alterity, an important ethical problem with our relations to "non-humans" in general is indeed that we make them into "non-humans": we only consider their reality and being in relation to us. Using the concept of hospitality, we can say that the ethical problem is not only that we are too little hospitable (we exclude many entities from the social) but also, and at the same time, we are *too* hospitable. We are used to drawing in all kinds of non-humans. We are used to dealing with the alien, with the mysterious, with the magical. But by inviting them into our linguistic and social *domus*, by including them into our family, we also may do violence to them by disregarding and disrespecting their difference and alterity, their otherness. The human sphere, by being hospitable, is also colonizing and violent. It sucks in all the entities into its gravitational field of meaning, language, and social relations, in which nothing and no-one is alien and nothing is truly other. By turning entities – including machines – into non-humans, they become fatally related to the human and are in danger of losing their otherness.

For our relations to others in general, this means that to propose an ethics of hospitality – as Derrida has done (Derrida 2000) – is a problematic solution to the problem of alterity. Of course on the one hand, it is good to be hospitable, it is good to accommodate someone. For Derrida, it is even a "law." But if, on the other hand, this hospitality implies that you force the other to become same, it is ethically problematic. In Derrida's concept of hospitality we find both sides. On the one hand, I should open my home to the foreigner as "the absolute, unknown" other – in other words I should respect the foreigner's alterity. On the other hand, Derrida writes that hospitality requires that I "*give place* to [foreigners], that I let them come, that I let them arrive, and take place in the place I offer them" (Derrida 2000, 25). But this place-giving may also reduce the alterity, and certainly the absolute alterity of the other (if such a thing exists at all), since by giving place I also include the foreigner (*xenos*) and alien into the home and the familiar. And as Derrida argues, naming and identifying is also problematic, especially if hospitality is not offered before asking the name (29). Once we ask the name, we already reduce the alterity of the other. If hospitality is the ethical way forward when it comes to our relations with human and non-human others, then it must be understood as inviting in the other but at the same time letting the other remain an other, respecting its otherness. But how can this be done, if at all?

To extend the investigation, in general but also as applied to the machine, I propose to further inquire into the phenomenology and epistemology of what happens in the encounter with the other, here the machine. What happens when we (do not) know and experience the machine as having alterity? How can we know it *as* alterity at all? How can we experience it *as* an Other?

A first guess is that for a machine to be experienced as an other, it needs to become present-at-hand, rather than ready-to-hand. Using Martin Heidegger's famous distinction from *Being and Time* (Heidegger 1927), one could say that the machine needs to be *vorhanden* rather than *zuhanden*. When we use the machine, it withdraws. We rely on it without thinking about it. If it has to become a quasi-other (Ihde 1990), then it needs to stand out from this background, it has to get into the foreground. However, if we say "quasi-other," then we have already named, categorized, identified, and defined. Then we have already started from the view that it is not really an other. Moreover, at first sight this view that the machine needs to be *vorhanden* suggests that it is *up to us* how we perceive the machine. We can either use it and then it disappears, or we can stand back and consider it *as* a machine. I will question this assumption. Moreover, if we consider it as a machine, that does not at all guarantee that we experience its alterity, or that we even open up to the possibility that potentially it might have alterity. We could take a theoretical perspective, and incorporate and domesticate it. We could also, as Heidegger suggested in his later work (see for instance Heidegger 1977), try to achieve a different relation to it and look upon it in a way we look at art. Perhaps this is a more promising route to recognizing its alterity.

The way forward I propose, then, which is also at the same time a way backward, is the following: we could turn to one of Heidegger's (and therefore Levinas's and Derrida's) sources of inspiration – early romanticism. In early romanticism, for instance in the works of Novalis and Hölderlin, we find an epistemology that (1) recognizes that there is always a space of unknowing and hence leaves some of the mystery and alterity intact and (2) recognizes that we do not have full control over the epistemic process: entities may show up or not, objects may conceal themselves or reveal themselves. For instance, Novalis emphasized a "sense for the particular, personal, unknown, mysterious, for that which is to be *revealed*, what . . . cannot be represented" (Novalis 1799, 162) which he also called the darkness and the night. And in Hölderlin's poetry gods flee and withdraw: it is not up to us whether they appear (or not). In other words, this epistemology assumes that we cannot know everything, and puts less emphasis on epistemic agency and more on what we may call epistemic patiency: on the knowledge and the experience that *happens* to me (or not).

Applied to the question concerning non-humans and alterity, this means that (1) even if we wanted to categorize, identify, and specify an entity, we should not assume that, even in our efforts of hospitality and domestication, we can fully know, fully nail down the entity in question and (2) that we should not assume that alterity itself is something that we can control and command; instead, it may appear and disappear, it may show or not show itself within a particular encounter and relation. The phenomenological and epistemological question is then not so much and not only "How can I know/grasp/control the object?" but rather "How does the (alien) object make itself known to me?"

Alterity (as alienness) then happens in the object-subject relation in the following ways: on the one hand, the object reveals itself to the subject, breaks in (sometimes uncalled); on the other hand, the subject alludes to and uses the object, but can never gain full epistemic control over it – some degree of alterity always remain. In the language of ontology and metaphysics, with Graham Harman (2011) we might add that what the object is exceeds our relation to it. This suggests that objects have properties we might not have access to. It makes a Kantian point about a noumenal world or a kind of noumenal sphere that clouds the object we do not have access to. But it is better to say that something is other not in its properties, but in the encounter, in the event, in the breaking in. Something interrupts. Alterity is a verb. It is about experiencing (*Erfahren*) and something happening to you, which can assume the dimension of a meeting or en-counter (*Wiederfahren*).

This point can further be supported by drawing on Levinas. Levinas argued that alterity does not depend on particular qualities that distinguish the other from me; that would nullify alterity (Levinas 1961, 194). The other should not be specified. He was opposed to ontology, since it reduces the other to the same:

> Western philosophy has most often been an ontology: a reduction of the other to the same by interposition of a middle and neutral term that ensures the comprehension of being. This primacy of the same was Socrates's teaching: to receive nothing of the Other but what is in me. . . . The neutralization of the other who becomes a theme or an object . . . is precisely his reduction to the same. . . . To know amounts to grasping being out of nothing or reducing it to nothing, removing from it its alterity. . . . For the things the work of ontology consists in apprehending the individual (which alone exists) not in its individuality but in its generality (of which alone there is science). (Levinas 1961 as found in Levinas 1999, 43–44)

A better epistemology, then, is not about what "is" and about things in their generality but about what happens between subject and object, or subject and subject, in an encounter in which there is no full knowledge, no transparency, no full comprehension. There is appearance and withdrawal, and there may be surprises. It is also more about an event and a process, and there is room for individuality and otherness. It is more like a drama with characters and events, with a plot and a narrative.[3]

Once we start from this epistemology, then, the problem regarding the phenomenology and ethics of alterity changes since the phenomenological-epistemological assumptions have been changed. Our responsibility to be hospitable is no longer only about achieving a kind of hospitality which actively tries to preserve alterity (even if it always also erodes it), or about an ethics which actively tries to give dues to entities such as machines. Instead, it is at least *also* about awaiting and trying to stay open to whatever may happen, to alterity that may (or may not) show itself. It is awaiting not so much knowing but the breakdown of what we know. Derrida's conception of hospitality may also help again here: this ethics is unconditional, not relying on identification, a name is not asked and reciprocity is not asked (Derrida 2000, 25). Full epistemic control of the entity is not demanded. This leads to an ethics that is also hospitable, but also more patient, and that accepts dependency and limits to control.

For "the machine," it means that we put less effort in trying to fix its ontological status and in trying to categorize and specify it – for example as a machine or another "genus" (to use a Levinasian term) – but that we also "let it be" and let it show itself, also in its potential alterity and difference, and not only in its sameness (same as other machines, same as other non-humans). It means that we remain open to new experiences that may *happen* when machines appear and disappear, with or without humans involved. It means that we must be prepared to accept that what we know – or think we know – about the technology, may crumble away.

Interestingly, this ethics also means that we do not try to identify beforehand whether what or who we encounter is human or non-human. Open to the alien, it asks for an opening up and a hospitality that is unconditional. Derrida writes about "who or what": "Let us say yes *to who or what turns up*, before any determination, before any anticipation, before any *identification*, whether or not it has to do with a foreigner, an immigrant, an invited guest, or an unexpected visitor, whether or not the new arrival is the citizen of another country, a human, animal, or divine creature, a living or dead thing, male or female" (Derrida 2000, 77). Thus, hospitality, according to Derrida, requires that we do not play the ontological game but give hospitality unconditionally. If applied to machines, it means that we first say yes to "who or what turns up," that we let it/her/him in and accommodate, without however immediately naming it and giving it a place in our ontological schemes.

What kind of human-machine relation is this? It seems that in so far as we have a *vorhanden* relation to the machine, the relation presupposes distance, a distance which may lead even to the modern skepticism that created the object-subject gap in the first place. Yet to respond to this objection by turning to praxis also seems to miss an aspects of the machine, the aspects indicated previously, aspects that are not entirely under our control. The epistemology and ethics proposed here, by contrast, gives priority to an attitude and engagement with the machine that can neither be reduced to a theoretical relation nor to praxis. First, in the event of the breaking in of the machine's (quasi-)alterity there might be no distance (if I am suddenly visited) or infinite distance (the absolute alien), but in any case there is no theoretical standing back. If and when alterity is experienced (at all), the machine resists our theoretical efforts. A response is called for. Second, in this kind of encounter there is not mere praxis, if praxis means that there is a high degree of agency and control over the machine. Hospitality means a non-theoretical inviting in, and does not (merely) use the other. It also means postponing naming; experience, engagement, and response take priority.

Indeed, in these experiences and encounters, the role of language is important. Language mediates our perception and *reception* of the other. If machines start using language (are given language), it is interesting to explore what this means for our reception and hosting of machines, and in general for our moral and social relation to machines and their potential alterity. How we address a robot, for instance, shapes how the robot appears to us (Coeckelbergh 2011); for instance it differs if we use "it," "you," or another term. This is already a kind of naming and categorizing. Whether or not we can experience *machine* alterity, alterity depends on the language we use – or indeed *cannot* use when we are suddenly surprised, confronted – when there is a meeting. If there is machine alterity at all (and this is true for alterity in general), then perhaps it (sur)faces most at times when language is not sufficient, when language fails – especially when it fails to grasp and control. When I speak about alterity I have already lost it. But there is also experience which comes before words, and perhaps in that moment alterity can be experienced.

Attending to language can also highlight other ethical aspects of (the phenomenology of) our relation to machines. For instance when we use the word "robot," we reduce the particular robot and the particular experience, event, and encounter to a category. Heeding what Derrida said about "animal" – "The animal, what a word!" (Derrida 2008, 23) – we could exclaim: "The machine: what a word!" or "Robot: what a word!"

What are the practical ethical implications of this discussion? Arguably, there are more urgent and important ethical problems, for instance with our relations to animals – and of course humans. Animals are not machines, they are another life form. They share life with us. They are more truly companions, who eat and share bread (food) with us. This calls for an ethics of sharing, which is more based on sameness. There is sameness between humans and animals. An ethics based on sameness is not necessarily bad at all, it is also good and also should have its place. Distinctions can and must be made. And there is also some sameness between humans and machines, perhaps, but it is of a different nature. There may be a lot *more* alterity in (our relation to) animals than in machines, a lot more in humans and animals that we cannot contain in any way, that we cannot identify, that must remain alien. Maybe machines lack the kind of alterity Levinas talks about, that alterity which is tied to transcendence. Maybe they have a different kind of alterity. Clearly there are differences, and we can and should talk about them. These are discussions worth having. But what has been said here also shows that it is problematic to only have an ethics based on sameness. It suggests, for instance, that we are currently too insensitive to problems that may be created by defining and categorizing before experience, instead of opening up to whatever may happen. For instance, using broad categories such as "machines" and "robots" – and indeed "humans" – is problematic. The sameness created here might stand in the way of an adequate analysis of the phenomenology of encounters with machines and of ethical problems in human-robot relations, for instance when the "human" in question is a particular vulnerable child and the "robot" is a very particular automated system used in a specific way.[4] An ethics of alterity and difference attends us to problems with generalization and naming. It suggest that we are too impatient when we immediately reach for epistemic closure, instead of waiting and experiencing what might happen (or not). If we exclude machines and other technologies from the ethical analysis, as Levinas and many others did, we miss the normative dimension of technologies and perhaps some of their "alterity" aspects. If we start with defining and fixing, if we start with domesticating rather than hosting, some of the meanings and aspects of the machine might already be lost. An ethics of alterity and hospitality may encourage us to face the encounter and help us to be a better host.

CONCLUSION

In this chapter I have discussed the possibility of machine alterity by asking how this alterity – and alterity in general – may appear and may be experienced, given that we domesticate, name, and categorize. I have suggested a different phenomenology and epistemology which accepts limits to epistemic control and is more sensitive to the process-like and dramatic dimension of the potential encounter with the machine and other "non-humans." I have then demonstrated the ethical consequences of this: I have argued that it is ethically problematic when we see "non-humans" as mere negatives of the human and if everything is domesticated and colonized by the human will and the human imagination. For machines, for instance, this implies that we *might* miss their alterity aspect. Yet based on an insight from early romanticism I have also criticized an assumption behind this claim about domestication: not everything can be known, grasped, and controlled anyway. Here the limitations of language also appear: to some aspects of entities we may only be able to "allude" to, to "suggest," etc. Some of the mystery remains and perhaps must remain. Either way, an ethics of openness, hospitality, and patience seems to be appropriate, when it comes to our relations to human and non-human others, where hospitality must be understood as at the same time drawing in and respecting alterity. A challenge for the future of the social and society is to keep questioning ways of thinking and doing that instrumentalize and disrespect otherness, that lead to a totality of sameness. A challenge for relating to machines and other technologies is to not arrive too soon at epistemic closure, but to create and keep open *other* ways to experience and make sense of them. A challenge for the future of relations to others is, among other things, to see what these virtues of openness, hospitality, and patience mean in the drama of concrete and various encounters, in the poiesis and praxis of relating and letting be.

NOTES

1. Note that the same can be said about humans: our moral relation to human others also depends on our encounters and linguistic and social relations. Humans are also often defaced and instrumentalized, which is of course ethically highly problematic. The hope here is that, if we re-think and achieve a different relation to non-humans, this will go hand in hand with, or at least also encourage, different, better ethical relations with humans. Clearly, the analysis also applies to humans; in this chapter I focus on machines given the theme of this book and given that much has already been said about humans.

2. The etymology of encounter relates the term to opposition or confrontation. At the very least, one can say that an encounter can have this aspect.

The other may "counter" me and vice versa. Interestingly, the "counter" in "encounter" is also related to the Latin *computare*: to compute. This is relevant to our discussion about machine alterity: the machine may count and (therefore) counter me. It may appear as an adversary and as an entity that challenges me.

3. This remark is inspired by Paul Ricoeur; however, here I will only suggest and not further develop this "narrative" aspect.

4. Currently, the author is involved in research on the ethical aspects of child-robot interaction in robot-assisted therapy for children with autism spectrum disorder (European research project DREAM).

REFERENCES

Coeckelbergh, Mark. 2009. "Robot Anthropology: Robots as Hermeneutic Devices for Defining the Human." Paper presented at *The European Conference on Computing and Philosophy* (E-CAP 2009), Universitat Autònoma De Barcelona, Spain, July 2–4.

Coeckelbergh, Mark. 2011. "You, Robot: On the Linguistic Construction of Artificial Others." *AI & SOCIETY* 26(1): 61–69.

Coeckelbergh, Mark. 2012. *Growing Moral Relations: Critique of Moral Status Ascription*. Houndmills and New York: Palgrave Macmillan.

Coeckelbergh, Mark. 2014. "Robotic Appearances and Forms of Life. A Phenomenological-Hermeneutical Approach to the Relation between Robotics and Culture." In *Robotics in Germany and Japan: Philosophical and Technical Perspectives*. Edited by Michael Funk and Bernhard Irrgang, Band 5: 59–68. Dresden Philosophy of Technology Perspectives. Frankfurt am Main: Peter Lang Edition.

Coeckelbergh, Mark and David J. Gunkel. 2014. "Facing Animals: A Relational, Other-Oriented Approach to Moral Standing." *Journal of Agricultural and Environmental Ethics* 27(5): 715–733.

Derrida, Jacques. 2000. *Of Hospitality*. Translated by Rachel Bowlby. Stanford: Stanford University Press.

Derrida, Jacques. 2008. *The Animal That Therefore I Am*. Edited by Marie-Louise Mallet. Translated by David Willis. New York: Fordham University Press.

Descartes, René. 2003 (1637). "Discourse on the Method of Rightly Conducting the Reason." In *Discourse on Method and Meditations*. Translated by Elizabeth Sanderson Haldane and George Robert Thomson Ross, 1–52. Mineola: Dover Publications.

Harman, Graham. 2011. "The Road to Objects." *Continent* 1(3): 171–179.

Heidegger, Martin. 1977. *The Question Concerning Technology*. Translated by William Lovitt. New York: Harper and Row.

Heidegger, Martin. 1996 (1927). *Being and Time; a Translation of Sein Und Zeit*. Translated by Joan Stambaugh. Albany: State University of New York Press.

Ihde, Don. 1990. *Technology and the Lifeworld: From Garden to Earth*. Bloomington: Indiana University Press.

Levinas, Emmanuel. 1991 (1961). *Totality and Infinity. An Essay on Exteriority*. Translated by Alphonso Lingis. Dordrecht: Kluwer Academic Publishers.

Levinas, Emmanuel. 1999 (1995). *Alterity and Transcendence*. Translated by Michael B. Smith. New York: Columbia University Press.

Novalis. 1997 (1799). *Philosophical Writings*. Edited by Margaret Mahony Stoljar. Albany: State University of New York Press.

Chapter 10

Another Alterity

Rethinking Ethics in the Face of the Machine

David J. Gunkel

In the face of the machine—whether an avatar, chatterbot, or social robot like Nao, Pepper, or Jibo—the question that immediately confronts us is, "Can such artifacts even have face?" Should technology, which has almost always been considered a tool or medium of human action (Heidegger 1977; Feenberg 1991), be considered another socially situated interactive subject? Is it possible for the machine, or even a particular machine, to be Other? In response to these questions, Sherry Turkle identifies what she perceives to be a potentially disturbing trend: "I find people willing to seriously consider robots not only as pets but as potential friends, confidants, and even romantic partners. We don't seem to care what their artificial intelligences 'know' or 'understand' of the human moments we might 'share' with them . . . the performance of connection seems connection enough" (Turkle 2011, 9). In the face of the machine, Turkle argues, we seem to be willing, all too willing, to consider these technological objects to be Other—not just a kind of surrogate pet but a close friend, personal confidant, and even paramour. According to Turkle's diagnosis, we are in danger of substituting the technological interface for the face-to-face encounters we used to have with other human beings. "Technology," she explains, "is seductive when what it offers meets our human vulnerabilities. And as it turns out, we are very vulnerable indeed. We are lonely but fearful of intimacy. Digital connections and the sociable robot may offer the illusion of companionship without the demands of friendship" (Turkle 2011, 1).

As one contemplates the contemporary situation—a situation where, for example, we share intimate details with so-called "friends" on Facebook, spend hours working and reworking the look and appearance of our avatars, or fall for and confide in what turn out to be mindless chatterbots—Turkle's argument definitely appears to be persuasive. But that is, perhaps, the problem. Turkle might have a handle on the individual psychological stressors, but the philosophical opportunities and challenges that appear in the face of this "machinic other" are much more disturbing and interesting. Consequently, the thesis of this chapter can be stated quite simply and directly: The problem with our socially situated and increasingly interactive devices is not that they substitute a machine interface for the face-to-face relationships we used to have with others. Instead, it is in the face of the machine that we are challenged to reexamine critically who or what is, can be, or should be Other. In other words, if the intimacy of network connections and social robots appear to threaten human sociality and communication, it is not simply because these mechanisms are being substituted for real human contact but because it is in facing the question of the alterity of the machine that we are forced to face-up to and reconsider the violent exclusions and defacings that have always been made in the face of others and in the name of ethics.

1. STANDARD OPERATING PRESUMPTIONS

Ethics, in both theory and practice, is an exclusive undertaking. In confronting and dealing with others we inevitably make a decision between "who" is morally significant and "what" remains a mere thing. As Jacques Derrida (2005, 80) explains, the difference between these two small words—"who" and "what"—makes a big difference, precisely because they parse the world into two camps—those Others who count as socially significant and those things which remain mere objects or instruments. These decisions (which are quite literally a cut or "*de-caedere*" in the fabric of being) are often accomplished and justified on the basis of intrinsic, ontological properties. "The standard approach to the justification of moral status is," Mark Coeckelbergh (2012, 13) explains, "to refer to one or more (intrinsic) properties of the entity in question, such as consciousness or the ability to suffer. If the entity has this property, this then warrants giving the entity a certain moral status." In this transaction, ontology precedes ethics. What something is governs how it is. Or as Luciano Floridi (2013, 116) describes it, "What the entity is determines the degree of moral value it enjoys, if any." According to this way of thinking—what one might call the standard operating procedure of moral consideration—the question concerning the moral status of others would need to be decided by first identifying which property or properties would be necessary and sufficient to have moral standing and then figuring out whether a particular entity (or class of entities) possesses this property or not. Deciding things in this fashion, although entirely reasonable and expedient, has at least four problems, all of which become increasingly evident and problematic in the face of the machine.

1.1 Substantive Problems

First, how does one ascertain which exact property or properties are necessary and sufficient for moral status? In other words, which one, or ones, count? The history of moral philosophy can, in fact, be read as something of an on-going debate and struggle over this matter with different properties vying for attention at different times. And in this process, many properties that at one time seemed both necessary and sufficient have turned out to be spurious, prejudicial or both. Take for example the following event recalled by Aldo Leopold (1966, 237) at the beginning of "The Land Ethic": "When god-like Odysseus, returned from the wars in Troy, he hanged all on one rope a dozen slave-girls of his household whom he suspected of misbehavior during his absence. This hanging involved no question of propriety. The girls were property. The disposal of property was then, as now, a matter of expediency, not of right and wrong." At the time Odysseus is reported to have done this, only male heads of the household were considered legitimate moral and legal subjects. Everything else—his women, his children, and his animals—were property that could be disposed of without any moral hesitation whatsoever. But from where we stand now, the property "male head of the household" is considered a spurious and rather prejudicial criteria for deciding who counts as a moral subject and what remains a mere object.

Similar problems are encounter with, for example, the faculty of reason, which is the property that eventually replaces prejudicial criteria like "male head of the household." When Immanuel Kant (1985, 17) defined morality as involving the rational determination of the will, non-human animals, which did not possess reason, are categorically excluded from moral consideration. The practical employment of reason does not concern animals and, when Kant does make mention of animality (*Tierheit*), he only uses it as a foil by which to define the boundaries of humanity proper. It is because the human being possesses reason, that he (and the human being, in this particular circumstance, was still principally understood to be male) is raised above the instinctual behavior of the brutes and able to act according to the principles of pure practical reason (Kant 1985, 63).

The property of reason, however, has been subsequently contested by efforts in animal rights philosophy, which begins, according to Peter Singer's analysis, with a critical intervention issued by Jeremy Bentham (2005, 283): "The question is not, 'Can they reason?' nor, 'Can they talk?' but 'Can they suffer?'" According to Singer, the morally relevant property is not speech or reason, which he believes would set the bar for moral inclusion too high, but sentience and the capability to suffer. In *Animal Liberation* (1975) and subsequent writings, Singer argues that any sentient entity, and thus any being that can suffer, has an interest in not suffering and therefore deserves to have that interest taken into account. "If a being suffers," Singer (1975, 9) argues, "there can be no moral justification for refusing to take that suffering into consideration. No matter what the nature of the being, the principle of equality requires that its suffering be counted equally with the like suffering of any other being."

This is, however, not the final word on the matter. One of the criticisms of animal rights philosophy, is that this development, for all its promise to intervene in the anthropocentric tradition and include others, still remains an exclusive and exclusionary practice. "If dominant forms of ethical theory—from Kantianism to care ethics to moral rights theory—are unwilling to make a place for animals within their scope of consideration," Matthew Calarco (2008, 126) argues, "it is clear that emerging theories of ethics that are more open and expansive with regard to animals are able to develop their positions only by making other, equally serious kinds of exclusions." Environmental ethics, for instance, has been critical of animal rights philosophy for organizing its moral innovations on a property (i.e. suffering) that includes some sentient creatures in the community of moral subjects while simultaneously justifying the exclusion of other kinds of "lower animals," plants, and the other entities that comprise the natural environment.

But even these efforts to open up and to expand the community of legitimate moral subjects has also (and not surprisingly) been criticized for instituting additional exclusions. "Even bioethics and environmental ethics," Floridi (2013, 64) argues, "fail to achieve a level of complete universality and impartiality, because they are still biased against what is inanimate, lifeless, intangible, abstract, engineered, artificial, synthetic, hybrid, or merely possible. Even land ethics is biased against technology and artifacts, for example. From their perspective, only what is intuitively alive deserves to be considered as a proper centre of moral claims, no matter how minimal, so a whole universe escapes their attention." Consequently, no matter what property (or properties) comes to be identified as morally significant, the choice of property remains contentious, debatable, and seemingly irresolvable. The problem, therefore, is not necessarily deciding which property or properties come to be selected as morally significant. The problem may be in this approach itself, which makes moral consideration dependent upon a prior determination of properties.

1.2 Terminological Troubles

Second, irrespective of which property (or set of properties) is selected, they each have terminological troubles insofar as things like rationality, consciousness, suffering, etc. mean different things to different people and seem to resist univocal definition. Consciousness, for example, is one of the properties that has often been cited as a necessary condition for moral subjectivity (Himma 2009, 19). But consciousness is persistently difficult to define or characterize. The problem, as Max Velmans (2000, 5) points out, is that this term unfortunately "means many different things to many different people, and no universally agreed core meaning exists." In fact, if there is any general agreement among philosophers, psychologists, cognitive scientists, neurobiologists, ethologists, AI researchers, and robotics engineers regarding consciousness, it is that there is little or no agreement when it comes to defining and characterizing the concept. As Rodney Brooks (2002, 194) admits, "we have no real operational definition of consciousness," and for that reason, "we are completely prescientific at this point about what consciousness is."

Although consciousness, as Anne Foerst remarks, is the secular and supposedly more "scientific" replacement for the occultish "soul" (Benford and Malartre 2007, 162), it appears to be just as much an occult property or what Daniel Dennett (1998, 150) calls an impenetrable "black box."

Other properties do not do much better. Suffering and the experience of pain—which is the property usually deployed in non-standard patient-oriented approaches like animal rights philosophy—is just as problematic, as Dennett cleverly demonstrates in the text, "Why You Cannot Make a Computer That Feels Pain." In this provocatively titled essay, Dennett imagines trying to disprove the standard argument for human (and animal) exceptionalism "by actually writing a pain program, or designing a pain-feeling robot" (Dennett 1998, 191). At the end of what turns out to be a rather protracted and detailed consideration of the problem—complete with detailed block diagrams and programming flowcharts—Dennett concludes that we cannot, in fact, make a computer that feels pain. But the reason for drawing this conclusion does not derive from what one might expect. According to Dennett, the reason you cannot make a computer that feels pain is not the result of some technological limitation with the mechanism or its programming. It is a product of the fact that we remain unable to decide what pain is in the first place. What Dennett demonstrates, therefore, is not that some workable concept of pain cannot come to be instantiated in the mechanism of a computer or a robot, either now or in the foreseeable future, but that the very concept of pain that would be instantiated is already arbitrary, inconclusive, and indeterminate. "There can," Dennett (1998, 228) writes in the conclusion to the essay, "be no true theory of pain, and so no computer or robot could instantiate the true theory of pain, which it would have to do to feel real pain." What Dennett proves, then, is not an inability to program a computer to "feel pain" but our initial and persistent inability to decide and adequately articulate what constitutes "pain" in the first place.

1.3 Epistemological Problems

As if responding to Dennett's challenge, engineers have, in fact, not only constructed mechanisms that synthesize believable emotional responses (Bates 1994; Blumberg, Todd and Maes 1996; Breazeal and Brooks 2004), like the dental-training robot Simroid that cries out in pain when students "hurt" it (Kokoro 2009), but also systems capable of evincing something that appears to be what we generally recognize as "pain." The interesting issue in these cases is determining whether this is in fact "real pain" or just a simulation. In other words, once the morally significant property or properties have been identified and defined, how can one be entirely certain that a particular entity possesses it, and actually possesses it instead of merely simulating it? Answering this question is difficult, especially because most of the properties that are considered morally relevant tend to be internal mental or subjective states that are not immediately accessible or directly observable. As Paul Churchland (1999, 67) famously asked, "How does one determine whether something other than oneself—an alien creature, a sophisticated robot, a socially active computer, or even another human—is really a thinking, feeling, conscious being; rather than, for example, an unconscious automaton whose behavior arises from something other than genuine mental states?" This is, of course, what philosophers commonly call "the problem of other minds," Though this problem is not necessarily intractable, as I think Steve Torrance (2013) has persuasively argued, the fact of the matter is we cannot, as Donna Haraway (2008, 226) describes it, "climb into the heads of others to get the full story from the inside."

Although "pain" is not the direct object of his analysis, the epistemological problem of distinguishing between the "real thing" and its mere simulation is illustrated by John Searle's "Chinese Room." This influential thought experiment, first introduced in 1980 with the essay "Minds, Brains, and Programs" and elaborated in subsequent publications, was offered as an argument against the claims of strong AI.

> Imagine a native English speaker who knows no Chinese locked in a room full of boxes of Chinese symbols (a data base) together with a book of instructions for manipulating the symbols (the program). Imagine that people outside the room send in other Chinese symbols which, unknown to the person in the room, are questions in Chinese (the input). And imagine that by following the instructions in the program the man in the room is able to pass out Chinese symbols which are correct answers to the questions (the output). The program enables the person in the room to pass the Turing Test for understanding Chinese but he does not understand a word of Chinese. (Searle 1999, 115)

The point of Searle's imaginative albeit ethnocentric illustration is quite simple—simulation is not the real thing. Merely shifting symbols around in a way that looks like linguistic understanding is not really an understanding of the language. A similar point has been made in the consideration of other properties, like sentience and the experience of pain. Even if, as J. Kevin O'Regan (2007, 332) writes, it were possible to design a robot that "screams and shows avoidance behavior, imitating in all respects what a human would do when in pain. . . . All this would not guarantee that to the robot, there was actually something it was like to have the pain. The robot might simply be going through the motions of manifesting its pain: perhaps it actually feels nothing at all." The problem exhibited by both examples, however, is not simply that there is a difference between simulation and the real thing. The problem is that we remain persistently unable to distinguish the one from the other in any way that would be considered entirely satisfactory.

1.4 Moral Problems

Finally, there are ethical problems with the properties approach. Any decision concerning qualifying properties is necessarily a normative procedure and an exercise of power over others. In making a determination about the criteria for moral inclusion, someone or some group universalizes their particular experience or situation and imposes this decision on others as the fundamental condition for moral consideration. It is, for example, because human beings experience suffering as both uncomfortable and a moral evil, that it is assumed that the same experience, or at least something substantially similar, in another entity, like an animal, would need to be evaluated and addressed in the same way. This is, for all its promise, still a form anthropocentric thinking and a variety of what the environmental ethicist Thomas Birch (1993, 315) calls "imperial power mongering," insofar as it evaluates the moral standing of others only to the extent that they are "just like us." This is precisely that kind of philosophical thinking that Emmanuel Levinas (1987, 43), criticized for "reducing to the same all that is opposed to it as other."

Consequently "the institution of any practice of any criterion of moral considerability," Birch (1993, 317) argues, "is an act of power over, and ultimately an act of violence toward, those others who turn out to fail the test of the criterion and are therefore not permitted to enjoy the membership benefits of the club of *consideranda*." In other words, every criteria of moral inclusion, every comprehensive list of qualifying properties, no matter how neutral, objective, or inclusive it appears, is an imposition of power insofar as it consists in the normalization of a particular value or set of values made by someone from a particular position of power. "The nub of the problem with granting or extending rights to others," Birch (1995, 39) concludes, "a problem which becomes pronounced when nature is the intended beneficiary, is that it presupposes the existence and the maintenance of a position of power from which to do the granting." The problem, then, is not only with the specific property or properties that come to be selected as the criteria of moral inclusion but also, and perhaps more so, the very act of selecting properties, which already empowers someone to make these decisions for others.

2. THINKING OTHERWISE

In response to these problems, philosophers—especially in the continental tradition—have advanced alternative approaches to deciding the question of moral standing that can be called, for lack of a better description, "thinking otherwise" (Gunkel 2007). This phrase signifies different ways to formulate the question concerning moral status that is open to and able to accommodate others—and other forms of otherness. And when it comes to thinking otherwise, perhaps no philosopher is better suited to the task than Emmanuel Levinas. Unlike a lot of what goes by the name of "moral philosophy," Levinasian ethics does not get caught up in efforts to determine ontological criteria for inclusion or exclusion but begins from the existential fact that we always and already find ourselves in situations facing and needing to respond to others. For Levinas, ethics transpires not in theorizing about the essential properties of others but in the very real "vulnerabilities," to use Turkle's terminology, that we already experience in the face of others. This change in perspective provides for a number of important innovations that, if pursued far enough, alter how we respond to, and in the face of, others.

2.1 Relatively Relational

According to this alternative way of thinking, moral status is determined and conferred not on the basis of subjective or internal properties decided in advance but according to objectively observable, extrinsic relationships. "Moral consideration," as Coeckelbergh (2010, 214) describes it, "is no longer seen as being 'intrinsic' to the entity: instead it is seen as something that is 'extrinsic': it is attributed to entities within social relations and within a social context." As we encounter and interact with other entities—whether they be another human person, an animal, the natural environment, or a domestic robot—this other is first and foremost experienced in relationship to us. The question of moral status, therefore, does not depend on and derive from what the other is in its essence but on how she/he/it (and the choice of pronoun here is part of the problem) stands in relationship to us and how we decide, in the face of the other (to use Levinasian terminology), to respond. Consequently, and contrary to Floridi's (2013, 116) description, what the entity is *does not* determine the degree of moral value it enjoys. Instead the exposure to the face of the Other, what Levinas calls "ethics," precedes and takes precedence over all these ontological machinations and determinations. And it is precisely for this reason, that Levinas (1969, 304) famously argued that "morality is not a branch of [applied] philosophy, but first philosophy" where "first" is understood in terms of both temporal sequence and status.

This shift in perspective—a shift that inverts the standard operating procedure by putting ethics before ontology—is not just a theoretical proposal; it has, in fact, been experimentally confirmed in a number of practical investigations with computers and robots. The computer as social actor (CSA) studies undertaken by Byron Reeves and Clifford Nass (1996), for example, demonstrated that human users will accord computers social standing similar to that of another human person and that this occurs as a product of the extrinsic social interaction, irrespective of the actual intrinsic properties (actually known or not) of the entities in question.

> Computers, in the way that they communicate, instruct, and take turns interacting, are close enough to human that they encourage social responses. The encouragement necessary for such a reaction need not be much. As long as there are some behaviors that suggest a social presence, people will respond accordingly. When it comes to being social, people are built to make the conservative error: When in doubt, treat it as human. Consequently, any medium that is close enough will get human treatment, even though people know it's foolish and even though they likely will deny it afterwards. (Reeves and Nass 1996, 22)

In the face of the machine, Reeves and Nass find, human test subjects treat the computer as another socially significant Other. In other words, a significant majority of test subjects respond to the computer as someone "who" counts as opposed to just another "what," and this occurs, they argue, as a product of the extrinsic social circumstances and often in direct opposition to the presumed ontological properties of the mechanism. These results have been verified in two recent studies with robots, one reported in the *International Journal of Social Robotics* (Rosenthal-von der Pütten et al. 2013) where researchers found that human subjects respond emotionally to robots and express empathic concern for machines irrespective of knowledge concerning the properties or inner workings of the device, and another that used physiological evidence, documented by electroencephalography, of the ability of humans to empathize with what appears to be "robot pain" (Suzuki et al. 2015). Although Levinas himself would probably not recognize it as such, what these studies demonstrate is precisely what he had advanced: the ethical response to the other precedes and even trumps decisions concerning ontological properties.

2.2 Radically Superficial

In this situation, the problems of other minds—the difficulty of knowing with any certitude whether the other who confronts us has a conscious mind or is capable of experiencing pain—is not some fundamental epistemological limitation that must be addressed and resolved prior to moral decision making. Levinasian philosophy, instead of being tripped up or derailed by this epistemological problem, immediately affirms and acknowledges it as the condition of possibility for ethics as such. Or, as Richard Cohen succinctly describes it, "not 'other minds,' mind you, but the 'face' of the other, and the faces of all others" (Cohen 2001, 336). In this way, then, Levinas provides for a seemingly more attentive and empirically grounded approach to the problem of other minds insofar as he explicitly acknowledges and endeavors to respond to and take responsibility for the original and irreducible difference of others instead of getting involved with and playing all kinds of speculative (and unfortunately wrongheaded) head games. "The ethical relationship," Levinas (1987, 56) writes, "is not grafted on to an antecedent relationship of cognition; it is a foundation and not a superstructure. . . . It is then more *cognitive* than cognition itself, and all objectivity must participate in it."

This means that the order of precedence in moral decision making can and perhaps should be reversed. Internal properties do not come first and then moral respect follows from this ontological fact. Instead the morally significant properties—those ontological criteria that we assume ground moral respect—are what Slavoj Žižek (2008, 209) terms "retroactively (presup)posited" as the result of and as justification for decisions made in the face of social interactions with others. In other words, we project the morally relevant properties onto or into those others who we have already decided to treat as being socially significant—those Others who are deemed to possess face, in Levinasian terminology. In social situations—in contending with the exteriority of others—we always and already decide between "who" counts as morally significant and "what" does not and then retroactively justify these actions by "finding" the properties that we believe motivated this decision making in the first place. Properties, therefore, are not the intrinsic *a prior* condition of possibility for moral standing. They are *a posteriori* products of extrinsic social interactions with and in the face of others.

Once again, this is not some theoretical formulation; it is practically the definition of machine intelligence. Although the phrase "artificial intelligence" is the product of an academic conference organized by John McCarthy at Dartmouth College in 1956, it is Alan Turing's 1950 paper and its "game of imitation," or what is now routinely called "the Turing Test," that defines and characterizes the field. Although Turing begins his essay by proposing to consider the question "Can machines think?" he immediately recognizes the difficulty with defining the subject "machine" and the property "think." For this reason, he proposes to pursue an alternative line of inquiry, one that can, as he describes it, be "expressed in relatively unambiguous words." "The new form of the problem can be described in terms of a game which we call the 'imitation game.' It is played with three people, a man (A), a woman (B), and an interrogator (C) who may be of either sex. The interrogator stays in a room apart from the other two. The object of the game for the interrogator is to determine which of the other two is the man and which is the woman" (Turing 1999, 37). This determination is to be made on the basis of simple questions and answers. The interrogator (C) asks both the man (A) and the woman (B) various questions, and based on their responses (and only their responses) to these inquiries tries to discern whether the respondent is a man or a woman. "In order that tone of voice may not help the interrogator," Turing (1999, 37–38) further stipulates, "the answers should be written, or better still, typewritten.

The ideal arrangement is to have a teleprinter communicating between the two rooms."

Turing then takes his thought experiment one step further. "We can now ask the question, 'What will happen when a machine takes the part of A in this game?' Will the interrogator decide wrongly as often when the game is played like this as he does when the game is played between a man and a woman? These questions replace our original, 'Can machines think'?" (Turing 1999, 38). In other words, if the man (A) in the game of imitation is replaced with a computer, would this device be able to respond to questions and simulate the activity of another person? If a computer is capable of successfully simulating a human being in communicative exchanges to such an extent that the interrogator cannot tell whether he is interacting with a machine or another person, then that machine would, Turing concludes, need to be considered "intelligent." Or in Žižek's terms, if the machine effectively passes for another human person in communicative interactions, the property of intelligence would be "retroactively (presup)posited" for that entity, and this is done irrespective of the actual internal states or operations of the other, which are, according to the stipulations of the test, unknown and hidden from view.

2.3 Literally Altruistic

Finally, because ethics transpires in the relationship with others or in the face of the other, decisions about moral standing can no longer be about the granting of rights to others. Instead, the other, first and foremost, questions my rights and challenges my solitude. According to Levinas (1969, 43), "The strangeness of the Other, his irreducibility to the I, to my thoughts and my possessions, is precisely accomplished as a calling into question of my spontaneity, as ethics." This interrupts and even reverses the power relationship enjoyed by previous forms of ethics. Here it is not a privileged group of insiders who then decide to extend rights to others, which is the standard model of all forms of moral inclusion or what Singer (1989, 148) calls a "liberation movement." Instead, the Other challenges and questions the rights and freedoms that I assume I already possess. The principal gesture, therefore, is not the conferring rights on others as a kind of benevolent gesture or even an act of compassion for others but deciding how to respond to the Other, who always and already places my rights and assumed privilege in question. Such an ethics is *altruistic* in the strict sense of the word. It is "of or to others."

For Levinas, however, this altruism appears to be limited. Whatever the import of his unique contribution, Other in Levinas is still unapologetically human. Although he is not the first to identify it, Jeffrey Nealon provides what is perhaps one of the most succinct descriptions of this problem:

> In thematizing response solely in terms of the human face and voice, it would seem that Levinas leaves untouched the oldest and perhaps most sinister unexamined privilege of the same: *anthropos* [*ἄνθρωπος*] and only *anthrōpos*, has *logos* [*λόγος*]; and as such, *anthropos* responds not to the barbarous or the inanimate, but only to those who qualify for the privilege of 'humanity', only those deemed to possess a face, only to those recognized to be living in the *logos*. (Nealon 1998, 71)

For Levinas, as for many of those who follow in the wake of his influence, the Other has been exclusively operationalized as another human subject. If, as Levinas argues, ethics precedes ontology, then in Levinas's own work anthropology and a certain brand of humanism precede ethics.[1]

This is not necessarily the only or even best possible outcome. In fact, Levinas can maintain this anthropocentrism only by turning "face" into a kind of ontological property and thereby undermining and even invalidating much of his own moral innovations. For others, like Mattew Calarco, this is not and should not be the final word on the matter: "Although Levinas himself is for the most part unabashedly and dogmatically anthropocentric, the underlying logic of his thought permits no such anthropocentrism. When read rigorously, the logic of Levinas's account of ethics does not allow for either of these two claims. In fact, as I shall argue, Levinas's ethical philosophy is, or at least should be, committed to a notion of universal ethical consideration, that is, an agnostic form of ethical consideration that has no *a priori* constraints or boundaries" (Calarco 2008, 55). In proposing this alternative reading, Calarco interprets Levinas against himself, arguing that the logic of Levinas's account is in fact richer and more radical than the limited interpretation the philosopher had initially provided for it. "If this is indeed the case," Calarco (2008, 55) argues, "that is, if it is the case that we do not know where the face begins and ends, where moral considerability begins and ends, then we are obligated to proceed from the possibility that anything might take on a face. And we are further obligated to hold this possibility permanently open" (2008, 55). This means, of course, that we would be obligated to seriously consider all kinds of others as Other, including other human persons, animals, the natural environment, artifacts, technologies, and robots. An "altruism" that limits in advance who can be Other is not, strictly speaking, altruistic.

3. WHAT OR WHO AND JIBO

We began with the question, "Can the machine have face?" This question, although seemingly direct and intuitive, might not only be the wrong question, but asking it might actually be an impediment to a solution. "There are," as Žižek (2006, 137) explains, "not only true or false solutions, there are also false questions. The task of philosophy is not to provide answers or solutions, but to submit to critical analysis the questions themselves, to make us see how the very way we perceive a problem is an obstacle to its solution." When we ask, "can the machine have face?" we already endorse two problematic assumptions. First, we use the word "machine" as a kind of umbrella term that gathers together all kinds of entities that are not necessarily the same or even similar. Like the word "animal," which gives Derrida (2008) considerable trouble, the word "machine" already makes questionable associations that reduce difference to the same and, at the same time, institutes other potentially problematic distinctions. We could, in fact, repurpose Derrida's exclamation about the word "animal" and apply it to this other word—this word that has been bestowed upon and used to name the other of the animal since at least the time of Descartes: "The machine, what a word! The machine is a word, it is an appellation that men have instituted, a name that they have given themselves the right and authority to give to others" (Derrida 2008, 23). The word "machine," therefore, does not name a neutral ontological category. It is already the product of a crucial decision that has been made in the face of others. In fact, it is precisely by conflating "animal" with "machine" in the hybrid term *bête-machine*, that modern philosophy, beginning with Descartes, succeeded in excluding both animals and machines from moral consideration.[2] Even before one makes a determination between "who" and "what," the machine—by its very name—is already (and along with its other, the animal) located on the side of "what."

Second, the verb "have," as in "have face" or "have a face," has the tendency to turn "face" into a possession and a property that belongs to someone or something. The form of the question, therefore, risks redeploying the properties approach to deciding moral standing in the process of trying to articulate an alternative. Instead of asking "Can the machine *have* face?" we should perhaps rework the form of the question: "What does it take for a machine to supervene and be revealed as Other in the Levinasian sense?" This question—which recognizes, following Coeckelbergh (in this volume), that "alterity is a verb"—no longer asks about "moral standing" in the strict sense of this term, since "standing" suggests that there is an ontological platform onto which morality would be mounted. It therefore comprises a more precise and properly *ethical* question—a question that remains open to others and other forms of otherness: "In what circumstance and under what conditions can a machine—a particular machine that appears here before me—take on face?"

In order to respond to this other question (a question that is otherwise and that can ask about others), we need to consider not "the machine" as a kind of general ontological category but a specific instance of an encounter with a particular entity. In July of 2014 the world got its first look at Jibo. Who or what is Jibo? That is an interesting and important question. In a promotional video that was designed to raise capital investment through pre-orders, social robotics pioneer Cynthia Breazeal introduced Jibo with the following explanation: "This is your car. This is your house. This is your toothbrush. These are your things. But these [and the camera zooms into a family photograph] are the *things* that matter. And somewhere in between is this guy. Introducing Jibo, the world's first family robot" (Jibo 2014). Whether explicitly recognized as such or not, this promotional video leverages Derrida's distinction between "who" and "what." On the side of "what" we have those things that are mere objects—our car, our house, and our toothbrush. According to the instrumental theory of technology, these things are mere instruments that do not have any independent moral status whatsoever (Lyotard 1984, 44). We might worry about the impact that the car's emissions has on the environment (or perhaps stated more precisely, on the health and well-being of the other human beings who share this planet with us), but the car itself is not a moral subject. On the other side there are, as the video describes it "those things that matter." These things are not things, strictly speaking, but are the other persons who count as socially and morally significant Others. Unlike the car, the house, or the toothbrush, these Others have moral status and can be benefited or harmed by our decisions and actions.

Jibo, we are told, occupies a place that is situated somewhere in between *what* are mere things and *who* really matters. Consequently, Jibo is not just another instrument, like our automobile or toothbrush. But he/she/it[3] is also not quite another member of the family pictured in the photograph. Jibo inhabits a place in between these two options. This is, it should be noted, not unprecedented. We are already familiar with other entities who/that occupy a similar ambivalent social position, like the family dog. In fact animals, which since the time of Descartes have been the other of the machine, provide a good precedent for understanding the opportunities and challenges of social robots, like Jibo. Some animals, like the domestic pigs that are raised for food, occupy the position of "what," being mere things that can be used and disposed of as we see fit. Other animals, like a pet dog, are closer to another person "who" counts as Other. They are named, occupy a place alongside us inside the house, and are considered by many to be "a member of the family" (see Gunkel and Coeckelbergh 2014).

As we have seen, we typically theorize and justify the decision between the *what* and the *who* on the basis of intrinsic properties. This approach puts ontology before ethics, whereby what an entity is determines how it comes to be treated. But this method, for all its expediency, also has considerable difficulties: (1) substantive problems with inconsistencies in the identification and selection of the qualifying property, (2) terminological troubles with the definition of the morally significant property, (3) epistemological complications with detecting and evaluating the presence of the property in another, and (4) moral concerns caused by the very effort to use this determination to justify extending moral standing to others. In fact, if we return to the example of animals, it seems very difficult to justify differentiating between the pig, which is a thing we raise and slaughter for food and other raw materials, and the dog, who is a member of the family, on the basis of ontological properties. In terms of all the usual criteria—consciousness, sentience, suffering, etc.—the pig and the dog seem to be virtually indistinguishable. Our moral theory might dictate strict ontological criteria for inclusion and exclusion, but our everyday practices seem to operate otherwise, proving George Orwell's *Animal Farm* (1945, 118) correct: "All animals are equal, but some animals are more equal."

Alternative approaches to making these decisions, like that developed by Levinas, recognize that who is or can be Other is much more complicated. The dog, for instance, occupies the place of an Other who counts, while the pig is excluded as a mere thing, not because of differences in their intrinsic properties, but because of the way these entities have been situated in relationship to us. One of these animals shares our home with us, is bestowed with a proper name, and is considered to have face, to use the Levinasian terminology. The other one does not. Jibo, like an animal, occupies an essentially undecidable position that is in between *who* and *what*. Whether Jibo is or is not an Other, therefore, is not something that will be decided in advance and on the basis of intrinsic properties; it will be negotiated and renegotiated again and again in the face of actual social circumstances. It will, in other words, be out of the actual social relationships we have with Jibo that one will decide whether he/she/it counts or not (and is therefore either a "s/he" or an "it").

Jibo, and other social robots like this, are not science fiction. They are already or will soon be in our lives and in our homes. And in the face of these socially situated and interactive entities, we are going to have to decide whether they are mere things like our car, our house, and our toothbrush; someone who matters like another member of the family; or something altogether different that is situated in between the one and the other. In whatever way this comes to be decided, however, these entities undoubtedly challenge our concept of ethics and the way we typically distinguish between *who* is to be considered Other and *what* is not. Although there are certainly good reasons to be concerned about how these technologies will be integrated into our lives and what the effect of that will be on us, this concern does not justify alarmist reactions and exclusions. We need, following the moral innovations of Levinas, to hold open the possibility that these devices might also implicate us in social relationships where they take on face. At the very least, ethics obligates us—and it does so in advance of knowing anything at all about the inner workings and ontological status of these other kinds of entities—to hold open the possibility that they might become Other. Turkle, therefore, is right about one thing, we are and we should be "willing to seriously consider robots not only as pets but as potential friends, confidants, and even romantic partners." But this is not a dangerous weakness or vulnerability to be avoided at all costs. It is ethics.

NOTES

1. For Derrida, the anthropocentrism of Levinasian philosophy constitutes cause for considerable concern: "In looking at the gaze of the other, Levinas says, one must forget the color of his eyes, in other words see the gaze, the face that gazes before seeing the visible eyes of the other. But when he reminds us that the 'best way of meeting the Other is not even to notice the color of his eyes', he is speaking of man, of one's fellow as man, kindred, brother; he thinks of the other man and this, for us, will later be revealed as a matter for serious concern" (Derrida 2008, 12). And what truly "concerns" Derrida is not just the way this anthropocentrism limits Levinas's philosophical innovations but the fact that it already makes exclusive decisions about the (im)possibility of an ethics that is open to and able to accommodate others, like non-human animals.

2. For Descartes, the human being was considered the sole creature capable of rational thought—the one entity able to say, and be certain in its saying, *cogito ergo sum*. Following from this, he had concluded that other entities, animals in particular, not only lacked reason but were nothing more than mindless robots that operated on the basis of pre-programmed instructions, like clockwork mechanisms. Conceptualized in this fashion, the animal and the machine were effectively indistinguishable and ontologically the same. (Descartes 1988, 44). Beginning with Descartes, then, the animal and machine share a common form of alterity that situates them as completely different from and distinctly other than human. For more on this association of the animal and the machine, see Derrida (2008) and Gunkel (2012).

3. The difficulty of deciding on a pronoun in this particular situation demonstrates the extent to which efforts to address this other kind of other not only strain against philosophical concepts but also the language available to express such concepts.

REFERENCES

Bates, J. 1994. The Role of Emotion in Believable Agents. *Communications of the ACM* 37: 122–125.

Benford, Gregory and Elisabeth Malartre. 2007. *Beyond Human: Living with Robots and Cyborgs*. New York: Tom Doherty.

Bentham, Jeremy. 1780. *An Introduction to the Principles of Morals and Legislation*. Edited by J. H. Burns and H. L. Hart. Oxford: Oxford University Press, 2005.

Birch, Thomas H. 1993. "Moral Considerability and Universal Consideration." *Environmental Ethics* 15: 313–332.

Birch, Thomas H. 1995. "The Incarnation of Wilderness: Wilderness Areas as Prisons." In *Postmodern Environmental Ethics*. Edited by Max Oelschlaeger, 137–162. Albany, NY: SUNY Press.

Blumberg, B., P. Todd and M. Maes. 1996. "No Bad Dogs: Ethological Lessons for Learning." In *Proceedings of the 4th International Conference on Simulation of Adaptive Behavior* (SAB96), 295–304. Cambridge, MA: MIT Press.

Breazeal, Cynthia and Rodney Brooks. 2004. "Robot Emotion: A Functional Perspective." In *Who Needs Emotions: The Brain Meets the Robot*. Edited by J. M. Fellous and M. Arbib, 271–310. Oxford: Oxford University Press.

Brooks, Rodney A. 2002. *Flesh and Machines: How Robots Will Change Us*. New York: Pantheon Books.

Calarco, Matthew. 2008. *Zoographies: The Question of the Animal from Heidegger to Derrida*. New York: Columbia University Press.

Churchland, Paul M. 1999. *Matter and Consciousness*. Cambridge, MIT Press.

Coeckelbergh, Mark. 2010. "Robot Rights? Towards a Social-Relational Justification of Moral Consideration." *Ethics and Information Technology* 12: 209–221.

Coeckelbergh, Mark. 2012. *Growing Moral Relations: Critique of Moral Status Ascription*. New York: Palgrave Macmillan.

Coeckelbergh, Mark and David J. Gunkel. 2014. "Facing Animals: A Relational, Other-Oriented Approach to Moral Standing." *Journal of Agricultural and Environmental Ethics* 27(5): 715–733.

Cohen, Richard A. 2001. *Ethics, Exegesis, and Philosophy: Interpretation After Levinas*. Cambridge: Cambridge University Press.

Dennett, Daniel C. 1998. *Brainstorms: Philosophical Essays on Mind and Psychology*. Cambridge, MA: MIT Press.

Derrida, Jacques. 2005. *Paper Machine*. Translated by R. Bowlby. Stanford, CA: Stanford University Press.

Derrida, Jacques. 2008. *The Animal That Therefore I Am*. Edited by Marie-Louise Mallet. Translated by David Wills. New York: Fordham University Press.

Descartes, René. 1988. "Discourse on the Method." *Selected Philosophical Writings*. Translated by J. Cottingham, R. Stoothoff and D. Murdoch. Cambridge: Cambridge University Press.

Feenberg, Andrew. 1991. *Critical Theory of Technology*. Oxford: Oxford University Press.

Floridi, Luciano. 2013. *The Ethics of Information*. Oxford: Oxford University Press.

Gunkel, David J. 2007. *Thinking Otherwise: Philosophy, Communication, Technology*. West Lafayette: Purdue University Press.

Gunkel, David J. 2012. *The Machine Question: Critical Perspectives on AI, Robots, and Ethics*. Cambridge, MA: MIT Press.

Haraway, Donna J. 2008. *When Species Meet*. Minneapolis, MN: University of Minnesota Press.

Heidegger, Martin. 1977. "The Question Concerning Technology." In *The Question Concerning Technology and Other Essays*. Translated by William Lovitt, 3–35. New York: Harper & Row.

Himma, Kenneth Einar. 2009. "Artificial Agency, Consciousness, and the Criteria for Moral Agency: What Properties Must an Artificial Agent Have to be a Moral Agent?" *Ethics and Information Technology* 11(1): 19–29.

Jibo. 2014. https://www.jibo.com

Kant, Immanuel. 1985. *Critique of Practical Reason*. Translated by Lewis White Beck. New York: Macmillan.

Kokoro, L. T. D. 2009. http://www.kokoro-dreams.co.jp/

Leopold, Aldo. 1966. *A Sand County Almanac*. Oxford: Oxford University Press.

Levinas, Emmanuel. 1969. *Totality and Infinity*. Translated by Alphoso Lingis. Pittsburgh, PA: Duquesne University Press.

Levinas, Emmanuel. 1987. *Collected Philosophical Papers*. Translated by Alphonso Lingis. Dordrecht: Martinus Nijhoff Publishers.

Lyotard, Jean-François. 1984. *The Postmodern Condition: A Report on Knowledge*. Translated by Geoff Bennington and Brian Massumi. Minneapolis, MN: University of Minnesota Press.

Nealon, Jeffrey. 1998. *Alterity Politics: Ethics and Performative Subjectivity*. Durham, NC: Duke University Press.

O'Regan, Kevin J. 2007. "How to Build Consciousness into a Robot: The Sensorimotor Approach." In *50 Years of Artificial Intelligence: Essays Dedicated to the 50th Anniversary of Artificial Intelligence*. Edited by M. Lungarella, F. Iida, J. Bongard and R. Pfeifer, 332–346. Berlin: Springer-Verlag.

Orwell, George. 1945. *Animal Farm*. New York: Harcourt Brace.

Reeves, Byron and Clifford Nass. 1996. *The Media Equation: How People Treat Computers, Television, and New Media Like Real People and Places*. Cambridge: Cambridge University Press.

Rosenthal-von der Pütten, Astrid M., Nicole C. Krämer, Laura Hoffmann, Sabrina Sobieraj and Sabrina C. Eimler. 2013. "An Experimental Study on Emotional Reactions Towards a Robot." *International Journal of Social Robotics* 5: 17–34.

Searle, John. 1980. "Minds, Brains, and Programs." *Behavioral and Brain Sciences* 3(3): 417–457.

Searle, John. 1999. "The Chinese Room." In *The MIT Encyclopedia of the Cognitive Sciences*. Edited by R. A. Wilson and F. Keil, 115–116. Cambridge, MA: MIT Press.

Singer, Peter. 1975. *Animal Liberation: A New Ethics for Our Treatment of Animals*. New York: New York Review of Books.

Singer, Peter. 1989. All animals are equal. In *Animal Rights and Human Obligations*. Edited by Tom Regan and Peter Singer, 148–162. New Jersey: Prentice-Hall.

Suzuki, Yutaka, Lisa Galli, Ayaka Ikeda, Shoji Itakura and Michiteru Kitazaki. 2015. "Measuring Empathy for Human and Robot Hand Pain Using Electroencephalography." *Scientific Reports* 5: 15924. http://www.nature.com/articles/srep15924

Torrance, Steve. 2013. "Artificial Consciousness and Artificial Ethics: Between Realism and Social Relationism." *Philosophy & Technology* 27(1): 9–29.

Turing, Alan M. 1999. "Computing Machinery and Intelligence." In *Computer Media and Communication.* Edited by P. A. Mayer, 37–58. Oxford: Oxford University Press.

Turkle, Sherry. 2011. *Alone Together: Why We Expect More from Technology and Less From Each Other*. New York: Basic Books.

Velmans, Max. 2000. *Understanding Consciousness*. London, UK: Routledge.

Žižek, Slavoj. 2006. "Philosophy, The 'Unknown Knowns', And the Public Use of Reason." *Topoi* 25: 137–142.

Žižek, Slavoj. 2008. *For They Know Not What They Do: Enjoyment as a Political Factor*. London: Verso.

Index

About the Contributors

Mark Coeckelbergh is a professor of philosophy of media and technology in the Philosophy Department of the University of Vienna and a part-time professor of technology and social responsibility at De Montfort University, UK. He is also the vice president of the Society for Philosophy and Technology. His publications include *Growing Moral Relations* (2012), *Human Being @ Risk* (2013), *Environmental Skill* (2015), *Money Machines* (2015), and numerous articles in the area of philosophy of technology.

Mira Fliescher is an assistant lecturer for modern and contemporary art at Karl Franzens-University Graz, Austria. She worked as a postdoctoral researcher at the Institute for Critical Theory at Zurich University of the Arts and at the DFG-doctoral research center *Visuality*, and as a lecturer in Cultural Studies and European Media Studies (University of Potsdam). Her publications are about alterity, the wit of art, the epistemologies of drawing, and art and art history. More information can be found at https://kfunigraz.academia.edu/mirafliescher.

Alexsandro Galeno is a teaching professor in the Department of Sociology at the Federal University of Rio Grande do Norte in Natal, Brazil. He is the author of several books like *Antonin Artaud. A revolta do anjo terrível* (2005); *Complexidade à Flor da Pele. Ensaioa sobre Ciência, cultura e comunicação* (2003); *Jornalismo e Literatura. A sedução da* *palavra* (2003); and *Ensaios Indisciplinados. Comunicação, cultura e arte* (2015).

David J. Gunkel is distinguished teaching professor in the Department of Communication at Northern Illinois University (USA). He is the author of five books, including *The Machine Question: Critical Perspectives on AI, Robots and Ethics* (2012), *Heidegger and the Media* (2014), and *Of Remixology: Ethics and Aesthetics after Remix* (2016). More information can be found at http://gunkelweb.com.

Ann Hetzel Gunkel is an associate professor of cultural studies and humanities at Columbia College Chicago. She is a two-time Fulbright recipient for both research (Germany, 1992) and teaching (Poland, 2011) and has won the Harmonia Research Grant from the Polish National Science Center (2012). Her research work in media, culture, and ethnic identity appears in the journals *Critical Studies in Media Communication, Popular Music and Society*, and *Polish American Studies*. More information can be found at http://www.annhetzelgunkel.com.

Maurício Liesen is a postdoctoral researcher at the University of São Paulo (Brazil), sponsored by the São Paulo Research Foundation (FAPESP). He is also a visiting lecturer at the University of Potsdam (Germany).

Ciro Marcondes Filho is a professor at the University of São Paulo (Brazil), creator of the *New Theory of Communication*, and head of FiloCom—The Centre for Philosophical Studies of Communication. He has published over forty-five books on journalism, mass media, cinema, and philosophy. His recent books include *The New Theory of Communication* (seven volumes), *The Face and the Machine* (winner of the Jabuti Award in 2014), and *The Dictionary of Communication.*

Dieter Mersch is the head of the Institute for Critical Theory at Zürich University of the Arts (Switzerland) and the author of several books, including *Was sich zeigt: Materialität, Präsenz, Ereignis* (2012), *Ereignis und Aura* (2012), *Posthermeneutik* (2010), and *Epistemologies of Aesthetics* (2014). For more information, see http://www.dieter-mersch.de.

Jörg Sternagel is an assistant professor of European media studies at the University of Potsdam. His research foci are: Theories of Alterity, Performativity, Image, Media Theory, and Philosophy of Existence. Recent publications include *Kraft der Alterität: Ethische und aisthetische Dimensionen des Performativen* (with Dieter Mersch and Lisa Stertz, 2015), *Paradoxalität des Medialen* (with Jan-Henrik Möller and Lenore Hipper, 2013), and *Acting and Performance in Moving Image Culture: Bodies, Screens, Renderings* (with Deborah Levitt and Dieter Mersch, 2012). More information can be found at http://www.emw.eu.

Tales Tomaz is an assistant professor at São Paulo Adventist University (UNASP), where he lectures on Cyberculture and Theories of Communication. He is a PhD candidate in Communication at the University of São Paulo (USP) with a scholarship from Capes-DAAD for a full year as visitor researcher in the Martin-Heidegger-Institut at the University of Wuppertal (Germany). He participates in FiloCom/USP (Center for Philosophical Studies in Communication) and in Cencib/PUC-SP (Interdisciplinary Center for Research in Communication and Cyberculture).

www.ingramcontent.com/pod-product-compliance
Ingram Content Group UK Ltd.
Pitfield, Milton Keynes, MK11 3LW, UK
UKHW040153091026
15587UKWH00018B/60

9 781783 488698